Up The Ante

A Magical Misfits Mystery

Lina Hansen

Literary Wanderlust | Denver, Colorado

Up The Ante is a work of fiction. Names, characters, places and incidents are either the product of the author's imagination or are used fictitiously, and any resemblance to actual persons, living or dead, business establishments, event or locales is entirely coincidental.

Published in the United States by Literary Wanderlust LLC, Denver, Colorado. www.LiteraryWanderlust.com

ISBN print: 978-1-956615-58-6
ISBN digital: 978-1-956615-59-3

Dedication

For Keith

Chapter One

D ried rose petals, brittle and pale, tumbled through the air. Most of them landed on the wooden floor; only the stupid ones caught in Tiddles's whiskers. With a growl, my geriatric kitty cat streaked from the living room as if possessed.

A soapy scent, rose blended with lavender, wafted from the open envelope in my hands and tickled my nostrils. "A-choo." I rubbed my itching nose.

Something small stuck inside the casing, so I gave the thing a healthy shake. A tiny bud in the same sorry state as the rest of the dead roses bounced onto my desk, but there was no letter, no note, nothing but my address and name on the envelope, written in a cramped hand as if begrudging the space.

The postmark was the only clue—Carnac in Brittany, France.

Carnac, with its Neolithic standing stones not unlike the ones in our village, but running in long rows, the *alignements*, rather than the somewhat gap-toothed circle that made up the Avebury henge.

Carnac, where my boyfriend Chris was supposed to finish a coding project, so I could join him on a much-needed vacation. Chris was also one of the few people who knew my magic manifested in rose petals.

A tendril of an unnamed emotion, which couldn't be fear—what was there to be afraid of?—wormed through my veins.

"Petty? See this? Doesn't look like Chris's handwriting to me." The tremolo in my voice betrayed the disquiet buzzing under my skin.

Accompanied by a flutter of her nubby leaves and a strong lemon scent, Petty, my zombie primrose familiar, landed her fake terracotta pot among the clutter on Aunt Eve's Victorian rolltop desk. She tilted her pot at my hand and then twirled her pink blossoms.

Duh. While she could "talk" and "see" things, my magical flower couldn't read. Her advice, however, was usually spot on.

"It's legible, that's why. As much as I love the man, his handwriting sucks."

Readability wasn't the only problem here. Somehow the writing gave off mean vibes. Chris didn't have a mean bone in his body. Instead, he had the kindest eyes. Night-dark, they crinkled at the corners every time he laughed. Which he did a lot. He also smelled wonderful, of fresh linen and incense. Not to forget his fingers, long and slender like a piano player's...

Get a grip.

Petty lifted her pot on a sparkly vortex and floated toward the envelope, sparks running along her leaves.

Until they winked out.

Usually, if she didn't like something, my primula drooped her foliage, but she wasn't doing that either. Instead, she sat on the floor, motionless, watching.

Deep inside, whatever nerve was in charge of creeping me out cranked up the ominous vibes. "Petty?"

My familiar shook herself with an angry rustle. The motion sent up a wave of a sharp, peppery aroma, a scent I had learned to fear during the latest spate of coven troubles.

"What's wrong? Ugh, sorry." Stupid of me, I should know better. Petty did a stellar job of getting her meaning across, but open questions weren't in the cards.

"Does the envelope bother you? Or is it about the petals?"

Petty banged her pot once, which meant yes. After a brief wait, she banged yes again, presumably answering my second question. It wasn't a no; of that I was sure. A no meant knocking twice in rapid succession.

"Surely, it can't be from Chris?" I winced. As a former language teacher, I should know better than to complicate matters with negatives.

Two rapid knocks on the wooden floor confirmed Petty understood just fine.

Rustle. With an angry kick from her pot, she swept aside some of the floral confetti, crumpling it. Then she shot upward, hovering in front of my face, the stalks swishing and waving in a magical storm. No translation app needed. I knew my plant. This morning's mail was unwelcome.

My nerves responsible for creepo alerts twitched in response, prompting a wave of exhaustion. With a groan triggered a long way down, I sank into the office chair.

What I needed was a break, not another busload of problems. Chasing after assorted killers, running a coven of wonky witches, and busting a curse had taken their toll. Not to forget my duties as the manager of the Witch's Retreat Bed and Breakfast. Though, courtesy of my housekeepers, the Simpkins sisters, my workload was reasonable by comparison. Still, I craved a holiday. I needed to hear from Chris that he was ready for me to join him. We wanted to hike, enjoy the sights, the good food and wine, and have a great time getting to know each other even better.

Instead, he'd gone incommunicado for over a day. And now this. Life wasn't fair.

I shook my head to clear my mind, and a strand of

strawberry hair slipped in front of my face, obscuring my vision. When I blew it away, it settled back into exactly the same place it had been in before. This time I left it and stared into space for quite a while.

Moping will get you nowhere.

As if she could read my thoughts, Petty twirled her blossoms.

I straightened. "Point taken. We have a problem, but we don't know what it might be. I'm keeping this stuff as evidence. No worries, I know better than to touch it with my bare hands." I fumbled for the pack of tissues lying next to the empty coffee mug.

Petty tilted her pot and, after a heart-stopping wait, rapped out a hesitant yes. This didn't mean we were in the clear and the petals or envelope were harmless, oh no. She—or whatever source she drew on—wasn't 100 percent reliable. She was a rookie, too. Like witch, like familiar. So far, we'd muddled through. However, magic was a risky business, and there was too much we didn't know.

I used the tissues to sweep the petals back into their envelope and hid it in a drawer. Better to keep the weird mail well away from Alma and Cecily's mops and squeegees—

Crap, what was that?

My fingers tingled. I jumped off my chair. The world tilted on its axis, only to steady itself with a crash. Though that might have been the office chair, now lying on the floor.

An ocean roared in my head, and the salty tang of seaweed ripened the air. A strange sensation—glee, or possibly spitefulness—blipped, only to vanish an instant later. The sensations, the ocean—all gone.

Petty, however, was still with me, her acrid pepper stink sharper than ever.

—

A knock sounded on the door.

"Eh, yes?" I croaked.

Another knock, this one more determined than the first.

I cleared my throat and stepped under the archway that led into the living room. "What's up? Come in."

The door opened, and Cecily Simpkins's solid figure appeared. Poodle-permed and pink-frocked, she would have cast well as Barbie's housekeeper.

"There's a gentleman outside what wants to see you."

I wiped my clammy hands on my jeans. "A guest? It's not even ten. He can't check in yet."

"Alma told him already. He doesn't want a room. He wants to talk to you." The frown on Cecily's forehead conveyed her opinion of our visitor. "Chappie needs to shift his posh car. He's blocking our entrance when the guests will soon want to leave."

Spooky petals or not, I knew my duties as owner and manager. Whoever the guy was, he needed my attention. Like now. "I'm coming."

"Good." Cecily withdrew her head but left the door open, a silent invitation to follow up on my promise.

I swung around and faced my familiar. "Got to sort something out, sweets."

Petty waved her leaves in acknowledgment.

Having slammed the drawer shut louder than necessary, I headed for the corridor. Halfway along, Alma, wearing a sapphire-blue house frock, her brunette hair poodle-permed like her sister's, pushed through the saloon-style swing doors that led to the kitchen.

"It's him."

I stopped. "Who? Chris?" No, it couldn't be him. The Simpkins sisters knew my lover since he'd slept over a few times. Cecily would have said something.

"Nah, that uncle of his," Alma confirmed.

My stomach did a belly-flop onto the coconut runner. "You're not talking about Bob Ignatius, are you?"

"I am. Nasty sort, that one."

Understatement of the year.

Top UK entrepreneur and scion of the witch hunters, Chris's uncle was one of the few people in the picture about the Avebury coven and its history. He knew back in the seventeenth century two factions of witches existed, the hotshot Whites and my lot, the lame-duck Reds. He knew the Whites disappeared through the circle, leaving their weaker cousins, my ancestors, at the mercy of the witch hunters, his and Chris's ancestors. To survive, the Reds disbanded and scattered. Throughout the centuries, the surviving Reds lost most of their lore and powers until Aunt Eve and Dot Wytchett called the last stragglers back to the henge.

That Ignatius knew about our existence was bad enough, but he was also the man whose lackeys caused the death of my parents, the same man who triggered Aunt Eve's lethal mistake. He had wanted first my aunt and now me to hex for him, to gain an edge over the competition. Did the tonker dare to renew his offer?

Fury seared my cheeks, but this time the sensation was uniquely mine. "I'll give him a piece of my mind."

Alma's "You do this" floated after me, but I was already halfway out of the door.

A sleek limousine that screamed serious money was indeed blocking the entrance. Next to it waited a middle-aged man with dark eyes and hair, the latter shot with silver. He radiated understated smartness—cream shirt, blue chinos, blue loafers, all of it expensive, none of it showy. He was neither too small nor too tall, and nothing in his features was in any way remarkable.

"Ah, Ms. Coldron, pleased to see you." His voice, cultured and pleasant, sent a chill down and up my spine.

"The answer is still no. Goodbye." Not the friendliest of

welcomes, but I doubted our visitor expected better.

"Charming as always." Ignatius's laugh grated on my teeth. Fortunately, it didn't last long. He licked his lips while his gaze, flickering from left to right, betrayed his wariness.

A faint buzz built in my head, as if my blood were fizzing and frothing. The fresh scent of moist grass and fertile soil filled my nostrils. Something soft welled from my palms, brushed past my fingers, and sailed to the ground.

In my peripheral vision, I spotted rose petals, smooth like a baby's cheek. My magic was manifesting. Blood red in the past, the petals had been bleached to powder pink by the curse that came close to nixing the Misfits. Despite the color change, my skylles seemed to work just fine.

I stepped up to Chris's uncle. To give him credit, he neither flinched nor twitched, his gaze never leaving my hands.

"If I were you, I'd give my next move some thought," I said.

Ignatius heaved a deep breath and looked up. "In case you wonder, I know the magic of the stronger witches manifests in flower parts such as petals. I just never saw it happen before."

For once, his voice didn't ooze fake charm. If anything, it trembled. "Yesterday, I received an unmarked envelope filled with ash. I had it tested. We're talking charred rose."

What the heck?

The fizzing eased off. A last petal joined the others, dotting the ground in a semicircle around my feet.

"Any message?"

"Nope, nothing. You're not responsible, are you?"

"No."

He nodded, but avoided my gaze. "I didn't expect you to be. Just wanted to be sure. Stuff gives me the willies, you know?"

Blimey, I never expected I could have something in common with the creep.

"Did it arrive by mail?"

"Yes." Ignatius withdrew a folded envelope from his shirt pocket. When I read the postmark, an imaginary icy finger traced my spine.

Carnac, France.

I sought Ignatius's dark gaze. "That's where Chris is." No doubt he knew that, but I said it anyway.

Chris's uncle flapped an impatient hand. "As you well know, my nephew doesn't pull such stunts. Is that all you can tell me?"

Oho, I wasn't quite ready to take the guy into my confidence. "Why would someone send you the ash of toasted roses?"

"It appears, Ms. Coldron, someone knows about witches, magic, and that my ancestors burned some of yours. And you haven't answered my question."

How observant of the little twerp. "Is that a problem for you?"

"Not necessarily. It depends on what this person wants. From both of us, perhaps?" He tilted his head.

To come clean on my mystery envelope or to keep my trap shut was the question. While I didn't need Mr. Slime Universe on the team, the fact we'd both received these odd missives was...well, unsettling wasn't quite the right word.

Ignatius heaved an exaggerated sigh. "In a way, I understand you. You fear I'm trying to trick you. Rest assured, I'm not. At least you've convinced me you aren't the sender of this deplorable mail. You might want to have a good mull over your next steps. Once you've decided, contact me. You have my number. In the meantime, as a sign of my good intentions, I suggest you check the news. My assistant is French. She pointed it out, that's how I know."

"I don't understand a word you're saying."

"It's about Chris. As I said, check the French news. Ours haven't picked it up yet."

He slid into the driver's seat and started the engine. An instant later, the window purred down. "Do me a favor and find out what stunt my least favorite nephew is running, will you? I hear you're quite good at clearing up murders."

Among a splatter of flying gravel, the car accelerated away.

Chapter Two

Lost for words, I stood in the car park, my thoughts spiraling down a black hole. Chris, I needed to get hold of Chris. Something was wrong and nothing else mattered. I fished for the smartphone in my pocket. The same moment my fingers touched the fake fur cover, a present from my cousin Daisy, I sensed someone's presence. I swung around and caught the gaze of a prim-looking sexagenarian, leaning over the immaculate fence of the house across the road.

"There you are," she said.

Urgh. Gloria Mornings—neighbor and passionate curtain twitcher—wasn't my favorite coven member. Like many of my fellow witches, she hadn't hexed anything yet. Unlike most of the others, she and her bosom buddy, Emma Bingham, took the paranormal failure personally and pinpointed me as the root of their problems, because I hadn't found the time to train anyone.

"Good morning, Gloria."

"Morning. Was that Bob Ignatius? What does he want?"

"Cause trouble?" I toyed with the fake fur covering my

phone. "Listen, I'm a bit busy—"

"You always are. I'd like to have a word."

I caught the groan by its tail and dragged it back before it could bolt. "If it's about the training, I know nothing much has happened—"

The expression on Gloria's thin face segued into grim. She opened her garden gate and strode across, muscling her way into my comfort zone. Whatever space remained, her oddly sweet floral scent filled in an instant.

"That's one way of putting it. Emma and I have always been clear on the matter. There's no point in us being here if we don't find our magic. Your aunt claimed once we gathered at the Avebury circle, things would fall into place for us, since the old stones would boost our powers."

Oh, Aunt Eve.

A quick look over my shoulder reassured me we were alone. If anyone was hiding behind their curtains, they'd notice nothing worse than an exchange between neighbors. "It doesn't seem to be quite so simple."

"Oh, really? If you'd only stop playing the amateur sleuth all the time and cared more for the coven, we would already know where we stand. Like you promised."

Whoa. Adrenaline spurted into my system. "Unfortunately, murders and magic seem to get mixed up in this place all the time, and I haven't worked out yet how to clone myself," I said. "I'll let you know when it happens. To be honest, I sometimes wonder if we have enough skylles left to boost them."

Gloria Mornings slumped. Her eyes, blue like spring flowers, lost their angry glitter. "You're telling me I might not have enough juju. At least you're honest. Unlike your aunt."

"Sorry. That was rather crass of me." I resisted a powerful urge to pat Mrs. Mornings's hand. Her pain was real and not something I could kiss and make good. Magic was important to the woman, and I'd just yanked the rug from under her feet.

I really need a holiday.

That was no excuse.

A painful longing showing in the woman's face, she tore her gaze from the rose petals littering the gravel. "Nah, you've confirmed my concerns. I guess I'd better face the truth."

"Don't give up yet. There are a few sections in Jenna's laundry list dealing with dormant magic and how to kick it into action."

"Kicking" and "action" didn't feature in the Wytchett grimoire. Stuffed to bursting point with words, the musty tome pontificated on the theoretical aspects of magic. On a plus point, the text in the laundry list was visible, quite unlike the juicy bits in the Coldron recipe book.

"I'll try my best, promised."

A chill draft that had nothing to do with summer gusted past, and Mrs. Mornings pulled her sage green cardigan closer to her bony frame. Since I wasn't wearing more than a T-shirt, I was left to shiver.

"I know you will," she said. "That's why I'm still here. I also understand you need a break. That's why I didn't want to..." She bit her lips.

"Hmm?"

"Oh, it's nothing. I think Emma's exaggerating as usual. She doesn't always think things through. Acts before she switches on her brain. I'll talk to her. You start the training when you're ready. And get some rest, will you? You're looking rather pinched recently." With a nod, she swung around and marched off. I wouldn't have been surprised had the colorful gnomes dotting her garden straightened and saluted their mistress, but that didn't happen.

What was bloody Mrs. Bingham up to now?

"Nothing like nosy neighbors, eh?" Alma marched up, banged open the lid of the wheelie bin, and emptied a load of rubbish into its stinky depths.

She was just as nosy as Mrs. Mornings and would no doubt want chapter and verse on the Ignatius skirmish. "The downside of small villages," I offered as a gambit to throw her off the scent.

Alma slammed the lid shut. "You're never alone, that's for sure. Apropos, what did the man have to say for himself?"

Straight to the point, in typical Simpkins sisters fashion. "Not here. Because—neighbors?"

She rattled her empty bucket. "True. It's cold as well. Let's go in."

We headed toward the house, with her whistling something off-tune and annoying, while I sifted through plausible diversion tactics that would allow me to escape and call Chris. The modest silver hatchback advertising Jamey's Knitwear wasn't quite it, but a bit of desperation went a long way.

I crunched to a halt and pointed at the vehicle. "Do we have a new sales rep?"

Word about our special discount for salespeople had spread, and these days rooms number one and two were booked solid. They were hard-working folks. If I could help, I would.

"Nah, it's only our Mrs. Shuttlecock. She's changed employers. Called me late last night. Fortunately, Randy Johnston canceled, so we could put her up in number two, what with Mrs. Saddler already in number one. Come on, it's nasty out today." Birkenstocks slapping, Alma strode ahead.

As a ruse, the car didn't work so well.

I arrived inside, and an insistent bleeping provided the much-needed inspiration. "Uh oh. Must fetch my washing from the dryer." A quick wave, and I was off.

The door to the utility room under the stairs locked, the washing machine rumbling away another load behind my back, I placed my call to Chris. The ringtone mocked my ear, but the call landed on his blasted voicemail, like it had done countless times before.

"You've reached Lentulus IT solutions. Please leave a number after the beep, and I'll call you back as soon as possible."

Beep.

"Chris, your uncle was here just now, blabbing about murders. What the blazes is going on? Call me ASAP, or I swear I'll catch the next ferry. Please, I'm worried."

I cut the call, willing Chris to pick up his phone and ring back.

But the gizmo remained obstinately silent.

French news, Ignatius said. My French sucked, but I should be able to figure out the gist of whatever was hitting the headlines. Le Monde was the only French daily I could remember, but nothing on their site caught my eye. Maybe something more local? I googled Carnac and anglais, because Chris was English after all, but apart from a recipe for Crème Anglaise, I drew a blank.

The washing machine I was leaning against shuddered into the spin cycle, sending vibrations into the ground under my feet. No intel on Chris on the national news surely implied nothing earth-shattering had happened, and bloody Bob Ignatius was up to his usual tricks.

Guessing wasn't the same as knowing.

The phone, which I'd placed on the top of the washer, jerked and skipped around like a neon-pink furry thing—

Neon-pink fake fur. Daisy. Where I scored mid-field at best with social media and stuff, my cousin was a true whiz.

I picked up the phone again and thumbed her number. This time, I connected after one ring.

"Myrtle," she squealed as if she'd hit the jackpot. "I was just about to call you."

Somehow, her comment didn't sound promising. "Uh, what's wrong?"

"Hello? I can hardly hear you. Are you in a thunderstorm?"

"No, it's the washer. I'm in the utility room."

"Oh-kaay. Do you think you could come to Mel's place? Like now? Hello, do you hear me?"

My mood, already dented by this morning's crap surprises,

took another hit. "I can hear you just fine. What's wrong?"

"Not exactly wrong. It's more like, weird? Look, Jenna wanted to see you, but you were busy with that tonker Ignatius, so she zipped over here. The Colonel has arrived as well. He couldn't reach you on the phone—did you turn your sound off? You always do that."

Weird was one way to describe what was going on.

"Stay where you are. I'm coming."

The washing machine decided this was a great time to trigger the final spin.

"Humming? Myrtle?"

"I said I'm coming. Don't go away," I yelled at the phone.

"Oh, I see. Don't worry, I'm going nowhere. Jenna's brought her latest creation. Raspberry yogurt tarts, they're absolutely delish," she raved.

Right. If the girls were scoffing cakes, Daisy's "weird" couldn't be a total disaster. Upon that cheerful note, I dashed for my cherry-red minivan.

—

The entrance of the Purple Emporium, Avebury's New Age shop, stood open, and the digital sound of waves sloshing over pebbles drifted out. A high, clear voice soared over the song of the ocean, intoning the same words over and over in a melodic mantra. The faint whiff of vanilla tea and incense pooling under the shop's purple awning amped up the soothing vibes, but right now Buddha himself couldn't have lowered my stress levels.

"Mel? It's me, Myrtle."

In a flutter of peach and saffron, Mel, the owner of all this exotic glory, appeared in the entrance. Chains of multi-colored beads twinkled in a sudden burst of sunlight that lit up her blonde hair like a halo.

"Three cheers to Daisy for tracking you down, girl. Come in and have some cake. Jenna's really surpassed herself."

I entered. "You better give me the bad news first, whatever

it is."

Mel's laugh, creamy and rich, made her chins wobble. "Girl, you radiate negativity. Stop fretting. It's not healthy."

So true. I heaved an incense-filled breath and let my gaze roam the shop.

The floral pom-poms in muted tones of duck-egg blue, sage, cream, and lime dangling from the ceiling drew the eye first. Underneath reigned a glorious chaos of shelves, open cupboards, baskets, and stands foaming with treasures, like the necklaces cascading from their hooks and the two walls crammed with medicinal herbs, spices, and teas. Crystals and stones, interspersed with an amazing collection of Tarot cards, old-fashioned CDs, and books on assorted aspects of the esoteric, filled the remaining walls.

I couldn't resist and dipped my hand into a bowl filled with shiny glass marbles, smooth yet firm under my fingers. My heartbeat slowed.

Light-footed like a dancer, Mel shimmied through the beaded curtain that separated the shop from the office beyond. Overflowing with yet more goodies, the office also sported two IKEA sofas and a chintz armchair, arranged around an engraved Moroccan brass tray groaning under a spread of tea things and assorted pastry.

"Ta-dah, I give you Myrtle Coldron. Am I good or what?" With that, Mel arranged herself in the armchair. It creaked once in protest, but then fell silent.

The sofas were occupied by the coven's finest, the bigger one by my local bestie, Jenna Wytchett, and our most senior member, Colonel Elmsworth. His dog Buster, the Chihuahua from hell, lay curled up at his feet. Not a single fang showed, which was amazing given the critter's rotten temper.

My cousin Daisy had draped her curves over the other sofa, playing with the fuzzy end of her auburn braid. Upon spotting me, she patted the empty seat next to her.

"Do sit."

I slumped onto the fake suede surface. "Is it that bad?"

Jenna's laugh always reminded me of Tinkerbell, as did her slim build and the pixie face smothered by a shock of maroon curls. "No, we figured you might need calories after your run-in with Ignatius. What did he want?"

Heavens, with my obsessing over Chris, I'd blanked the petals. "He received a strange letter from Carnac. Uh, strike that. Not a letter but an envelope filled with rose ash. As did I. Only mine contained dried rose petals instead of ash."

The room fell silent. Four pairs of eyes stared at me.

Until Jenna bounced from the sofa. "What? I got one as well. This morning, actually."

Fear shivered down my spine. "No way."

"Way," Jenna said. "It bothered me rotten. That's why I wanted to have a chat. Here, look at this."

She handed me a bulging envelope. I took a peek. Yup, more petals. While they matched the ones I received in the mail, there was no address on the envelope, no stamp, and no postmark. Instead, it bore Jenna's name in printed letters and nothing else.

"It must've been delivered by hand."

"Looks like it," Jenna said.

Yet another blasted complication. "Did you touch them? I used tissues, and they still made my fingers tingle. Plus, I suffered an odd vision."

"Amazeboggling," Jenna said. "What sort of vision? I noticed nothing."

"Think ocean sounds and smells. And the ground under my feet slipped. Didn't last long, though." I wasn't quite ready to mention the emotions carried along by that ocean breeze.

Jenna's gaze slid to the envelope in my hand. "Care to try mine?"

"Not sure I want to repeat the experience."

"Here, this'll help." Mel pushed a linen napkin at me.

The napkin wrapped around my digit, I tapped Jenna's

petals. When the visions failed to materialize, I tapped them again. Emboldened, I repeated the exercise without protection. "Nada. Zilch. No ocean sensations, no wobblies, thank the heavens."

I mopped my sweaty brow with the napkin.

Colonel Elmsworth, sipping his tea, pulled a face and returned his mug to the tray. "Carnac isn't far from the Atlantic, correct?"

I handed the napkin back to its owner. "Yes. Golfe du Morbihan."

Once again, the room fell silent.

"But what is this supposed to mean?" Daisy asked eventually.

"It means we have a moron among us," Elmsworth said, his sandy-gray mustache bristling. "Who outside the coven would know magic sometimes manifests in petals and that Myrtle's partner is in Carnac?"

"Ignatius, for one. That's why he came to me."

"Why would any of us prod the resident villain?" Mel snapped up a tart. Small and round, the raspberry-topped delight snuggled into the palm of her hand.

I couldn't stop my greedy fingers from picking up the pastry's twin. "I don't have the faintest clue. Why someone would deliver one letter by hand and mail the other ones from Carnac is even more mind-boggling. I mean, it's not exactly around the corner, is it?"

"Chris is." Mel bit into her cake.

"Yes, but why would he bother? And there's another problem."

"He still hasn't called?" Daisy asked.

"No, nor has he answered my messages."

The Colonel smiled at me. "Keep calm and ring on. He's probably knee-deep in coding and doesn't check his phone. As for the letters, I'm convinced they're nothing but a brain-dead attempt to rattle your cage, so you don't leave but start the training. Let me have a word with Emma. I wouldn't be surprised in the slightest if she's behind this idiocy."

That might have been what Mrs. Mornings meant when she claimed Mrs. Bingham was "exaggerating." There was only one fly in the Colonel's ointment. To the best of my knowledge, Mrs. Bingham hadn't left the country. Besides, she didn't have any magic, so how could her petals induce visions?

Hmm, I might have overreacted. Perhaps Emma Bingham had a friend over there? Or we had the wrong end of the stick?

In any case, the blasted missives had slipped on my list of priorities.

"It gets worse. Ignatius jabbered about Chris being involved in a murder and wanted me to find out what was going on. Something concerning the French news. I found nothing, the reason I called you, Daise."

Once more, a heavy silence grew in the room. I looked at my hand, still holding the cake, and took a bite, but somehow the fun had gone out of eating.

Daisy stirred. "Let me check."

She pecked around on her phone. While I was still searching for a place to leave my cake, my cousin let out a yelp.

"Oh my," she said. When she looked up, her pupils had dilated into inky pools. "That's like, not good?"

The tart dropped from my hands and landed on the floor. With a happy yip, Buster pounced.

"Show me."

She handed over her phone.

Panic clogging my throat, I stared at the screen, displaying a newspaper article. Most of it was behind the paywall, but not the picture. It displayed concrete steps leading up from a beach. Huddled by the steps lay a human shape, twisted like a rag doll. Whoever had taken the photo hadn't been close, so the image appeared somewhat blurred. The reason for the distance was all too apparent—a cordon of police officers blocking off the scene.

My brain pushed every single panic button at once. A

nauseating chaos of images exploded, and a shrill warning sounded in my mind.

Was this the reason Chris didn't respond to my calls? Because he was...dead? Life couldn't be so cruel.

For a vast part of humanity, it is. Why should you be spared?

I called my idiotic imagination to order, but the nausea refused to budge.

Then I noticed.

Chris always wore black. He wouldn't be seen dead in a cherry-red sweater, cardigan, or whatever the corpse was wearing.

Relief washed over me, making me dizzy. "Oh gosh, for a moment I thought it was him."

"Him what?" Daisy asked.

"Chris. But he doesn't wear red, ever."

Comprehension dawned in her brown eyes. "Oops, didn't want to shock you, Myr, sorry. He'll be fine. Though...oh, see for yourself." She swiped to another image and tapped the screen with one glittery fingernail. "This guy here, the one to the right. He looks a lot like Chris, doesn't he? And the caption below mentions a Brit being questioned in the context of a suspicious death. Myr, surely they won't think your Chris is a killer?"

Chapter Three

To imagine Chris would ever murder anyone was beyond ludicrous. However, Daisy was unfortunately correct. Not only did the man in the picture look like my boyfriend, but he also wore black and a familiar scowl.

One left, one right, Mel and Jenna peeked over my shoulder.

"Yep," Mel said. "It's him. Oh dear. Would you like some chamomile and lavender tea, girl? It'll do wonders for your nerves."

"I'm fine." I wasn't, not really, but chamomile tea evoked nasty memories of upset tummies and summer weekends spent in bed as a kid.

The Colonel, who had been wiping cream off Buster's muzzle, looked up. "Call your police detective friend. She'll know what to do."

Now, that suggestion carried a lot more promise. I'd met Sergeant Sarah Widdlethorpe from the Wiltshire CID during the investigation into my aunt's death, and she'd fast become another great friend.

I whipped out my phone. More often than not, my calls crash-landed in Sarah's voicemail. No surprise there—she was one busy woman. This time, however, the Fates must have taken mercy on me. One ringtone was all it took.

"Please, don't let it be another killer lurking behind the nearest dung heap."

Police officers have a bizarre sense of humor. "Rejoice, no one has died yet. At least not here in the village."

A groan fanned my ears. "You're not helping with my mood."

"Seriously, it's not your problem. But I need your help."

"Something's wrong. I can hear it in your voice."

"Eh, yes, you can say that."

"Shoot."

I took that to mean she wasn't referring to target practice. "It's Chris. He's—"

"Didn't he cross the Channel?"

"Yes, we were planning to spend a week in Carnac."

"Planning to?"

"Well, that's the point. He had a job to finish first, so I've been on standby. However, now he seems to have run into...eh, I don't know exactly what happened. I only just saw the news. It appears there's a suspicious death in Carnac, and the French police seem to be breathing down his neck."

A deafening silence filled my ears. Then she asked, "Are you telling me your sweetheart is a murder suspect?"

"Told you I don't know." There was enough acid in my voice to dissolve a whole plateful of cakes.

"Relax, will you? What do you know?"

"Very little. All I have is a news article mentioning a suspicious death and a Brit. We're agreed Chris must be the man in the picture."

"He might be a witness. Did you get a chance to talk to him?"

"No, he's not returning my calls. Maybe the idiots...I mean local law enforcement have taken him into custody."

"Thanks a bunch for the comment about the idiots."

"Hey, if the shoe fits."

"They're doing their job. Which brings us to the point. I'd love to help, but I'm not sure how."

I had to be careful here. Sarah was every inch a copper, and she wouldn't break a single rule, not even for a friend. "Could you find out how serious this is? Maybe you're right, and the situation isn't quite as dire as it looks. Until I get hold of him, I'm stuck."

"Hmm, yes, I'd be worried too if I were in your shoes. I can sniff around and see if les flics are willing to communicate. Don't expect miracles, though. Your Chris doesn't exactly top Europol's most-wanted list, so I don't have much leverage."

An avalanche rumbled off my heart. "I understand. And, Sarah? Thanks. You rock."

"Heh, I was just congratulating myself how quiet my life was recently and that my inbox was only overflowing, not exploding."

"Ack, I'm sorry. I owe you big time."

"Actually, it's the other way around. You and your hunches are helpful. Don't freak if I don't call back today, okay? It might take a moment to storm la Bastille. Au revoir." With that, she was gone.

I stared at my phone. Could it be the professional congratulated the amateur sleuth, aka myself, on her meddling?

"And?" Mel asked.

"She'll talk to French law enforcement."

"Good." Elmsworth rose and shoved a bunch of soiled tissues into the pockets of his corduroy trousers. "Now, that's taken care of. Nothing more you can do now. As for those letters, I'll have a word with Emma. In fact, I better do it straightaway. Thank you, Mel and Jenna, for the tea and cakes."

He and Buster left. I followed soon after having taken no more than that a single bite of my cake—a personal best.

—

Relief didn't last long. On the drive home, my imagination rapid-fired volleys of disaster images—Chris arrested, Chris convicted of murder, Chris languishing in jail—so it was a stroke of good luck I made it home in one piece. Fate being a bitch, two sales ladies were heading for their cars when I rolled in. They stopped and waved.

Rats. I would have to engage in customer liaison.

With a groan stuck deep in my throat, I left my transport and stepped up. "Hiya, how's things?"

"Your place is so restful," Flora Saddler said. Tall and lean, with the figure of a runner, she sprinted through life. She also sold and serviced gym equipment. No surprise there. That she stayed with us despite her cat allergy came as more of a surprise, but Tiddles wasn't keen on our guests and kept her distance. Most of the time.

"True," Iris Shuttlecock said. "You and your housekeepers are totally awesome, luv." Despite living in the UK long enough to take on local slang like "luv," she still sounded American to me.

"What happened to Randy?" she asked. "I hear I snagged his room. Must buy him a beer or something."

"He canceled," I said. "Don't know why."

"How odd," Mrs. Saddler said. "You can set the clock by him."

"True. I hope he's okay." I shifted from one foot to another, edgier than a cat in a room full of mutant monster mice. Small talk mustn't be rushed, but luckily both women drove off soon after, saluting me with their horns.

Once my guests had left, I pottered through the rest of my day, but Chris never called. Neither did Sarah, but she at least warned me. To keep myself occupied, I contacted a friendly solicitor and asked him to source a colleague in France. Until Chris or Sarah got in touch, I couldn't do more.

In the meantime, my rampant imagination sparked yet more nasty images. Chris handcuffed, his cute dark locks all messed up, his face haggard and gray. Chris in a cell, locked up for life because he couldn't prove his innocence—

Stop this crap. Stop it right now.

I'd googled that one already. More than once. Even in France, people were considered innocent until proven guilty, not the other way around, as popular opinion would have it.

Still, a stressed cat had nothing on me when I paced from the living room into the den and back again.

Apropos cat. "Tiddles?"

The wool in the knitting basket shifted, and a mottled head with pointy ears popped up.

"I need fur therapy. Please?"

She blinked and sank back into my aunt's unfinished knitwork. There was me, put in my place for daring to disturb the resident feline without a food offering.

Petty, dozing in her pot, wriggled her leaves once.

As I continued my pacing, a pale pink petal materialized in midair and sailed to the floor.

I swung around and faced the bookshelf, where the recipe book quivered, ready to drop into my hand. I'd placed the tome out of reach of the Simpkins sisters' dusters, right next to the urn containing auntie's ashes, but the blasted grimoire had its own mind and kept changing positions.

Smooth and warm, it lay in my hand. I opened a page at random. One of the empty ones, of course. They weren't truly empty, as we now knew. An ancestor, Lily Coldron, had penned lots of instructions back in the sixteen hundreds and hexed them to become invisible. It had taken Daisy's and my combined efforts to work that out.

Ah, Daisy. Should I wake her? She was sleeping upstairs in her room, having assured me many times she'd be there for me any time I felt like hollering.

Nah, she was running a shop, and she needed her sleep. I

had the Simpkins sisters, which meant I could rest any time I felt like it.

That was assuming I'd be able to power down.

Anyway, what would seventeenth-century Lily have known about my problems in the here and now? In a surge of desperation, I placed my hand on the blotchy page.

"I want Chris to call."

Carried by the odor of wet soil, a whole flurry of powder-pink petals materialized in midair and landed on the coffee table.

I waited.

Nothing.

Duh, what do you expect? This is a grimoire, not a wishing well.

And while I might be lots of things, I was no fairy.

The recipe book shut and placed aside, I fired up my tablet. With the paranormal on the blink, I might as well try the amateur sleuth angle. Writing down the facts helped to clear my mind, and boy, that mind needed clearing.

We knew so little.

Three letters, two mailed from Carnac, one hand-delivered in Avebury. Two letters stuffed with rose petals, one filled with rose ash. No messages of any sort, only two handwritten addresses. A man dead in distant France and my partner somehow entangled in the investigation.

I doodled a vortex and labeled it "Carnac."

My inner Watson agreed the town linked the letters to Chris's troubles. Which probably meant the whole stupid petal business had little to do with delayed hexing lessons. Mrs. Mornings had a point when she said both she and Emma Bingham had been clear about their motivations. No need to send moronic letters.

I doodled a garden gnome.

But the Colonel was right to suspect an insider. Who else would be aware of the coven's existence and Ignatius's nasty ancestry?

I could hear Sarah's voice in my head. Too much speculation. Look at the facts, it told me.

Facts I could do.

That our witchy ancestors, the Earth Wardens, dwelled at the henge in the sixteen hundreds was an accepted part of local history, though few people believed in their magic. Quite a few of the savvier locals had figured out we newbies were descended from the Wardens. Would anyone believe the bit about the magic also applied to us?

They might. If they were open-minded. Like the Wiccans, Druids, and other Pagans flocking to the henge.

Yet why would any of them send us these letters? Letters even a hard-boiled businessperson like Bob Ignatius seemed to consider a threat? How did Chris's troubles fit into the picture? How could someone be both in Carnac and here in Avebury? Our magic was pretty kaput. Teleporting was out of the question.

I doodled a big question mark with wriggly lines radiating from it.

Unless Chris did send the letters after all. But every fiber in my body revolted against the thought.

Somewhere in the distance, the clock in the church tower started the countdown to midnight. The last bong vibrated in my ear—and was chased away by an eerie noise blasting from my phone.

I jumped back. My heart beat double-time into my mouth.

The stupid phone continued the racket—a discordant mixture of moaning wind and something that sounded like a siren in the distance. What a crap ringtone.

A hoarse male voice said, "Billie."

On second thought, was it a ringtone? "I'm...I'm Myrtle," I said. "Hello?"

When a woman started singing, I figured it out. My phone was playing Billie Eilish's "Bury a Friend." Not one of my favorites, and I could swear I never programmed that one.

The phone in my hand, I checked the caller ID. My heart did a skip and hop, this time with joy. I thumbed the call open.

"Chris, thank the heavens."

His sigh gusted into my ear. "Sorry, sorry, sorry. I know this is a crazy time to call. Couldn't make it any earlier, though. I'm...I'm having trouble with the local police. I was stuck at the cop shop, which took absolutely ages. Plus, I wanted to know where this was headed before getting in touch."

"I saw the news. They can't seriously believe—"

"Officially, I'm helping with inquiries. They couldn't take me into custody, since there's no evidence I killed the poor guy, no matter how hard this idiotic copper tried to make me confess the crime. He relished telling me I'd go straight to prison as soon as they have their evidence."

"Rubbish. Even if push came to shove, surely they have bail in France?"

"They do, but it's not for me. As a foreigner, I'm a flight risk. Bloody hell, I'm not a criminal." Chris's voice rose to a shout. If I needed any further clues concerning his state of mind, here it was. The man didn't do panic.

Tears stung my eyes. I wiped them away with the heel of my hand. One of us needed to stay in control. "Of course not. Shall I come?"

"No. I mean, not yet. As much as I'd love to have you here, things are too much up in the air. Who wants to shack up with a criminal, anyway?"

While I couldn't blame him, his black mood wasn't helping. I rubbed my sternum, as if that would coax whatever healing magic I might possess. "Chris, calm down. Can you do me a favor and tell me what happened? All I know is someone's dead, and you're a suspect."

"Do you seriously want to hear the sordid—"

"Hey, I'm trying to help."

"Sorry. Of course, you are, dear," he said in a voice filled with so much longing, I would have teleported to France there and then. If only I could.

"It's okay," I whispered, unable to ramp up the volume.

"I'm getting to the point where I can tell the story in my sleep. After finishing the job—yes, I'm done, but it won't do us any good—I was knackered and went to bed with the chickens. The next morning, I woke up with this mad urge to move. So, I went jogging. Got this new smartwatch I wanted to put to the test—your fault, you know?"

Ah, an attempt at banter. Much better. "Oi, since when are your techie gadgets my responsibility?"

"May I remind madam of certain derogatory comments about my old watch? Comments like 'Klingon communicator' or 'Mafia bling'?"

Now he sounded like himself. "That clunky thing."

"Yeah, it was rather heavy. The new one is cool. It doesn't brew coffee or change tires, but apart from that, there's not much it can't do."

I smiled. Chris's enthusiasm was infectious.

"Pity no one saw me when I left my auberge."

"What time was it?"

"Quarter to six."

"Ouch. That *is* early."

"As I said, I was stir-crazy. The sun hadn't risen yet, but it was light enough. Originally, I planned to go east, but weirdly enough, I somehow turned the other way, arrived at the beach—and bingo, there he was."

As if touched by a dead finger, the fine hairs on my neck rose. "The...corpse?"

"Who else? Napoleon?" The edge was back in Chris's voice.

Proceed with caution. The man had every right to be jumpy.

Another sigh puffed into my ear. "Oh, blast. Didn't mean to take it out on you."

"It's all right. What exactly did you see? I'm asking on behalf of Sarah, by the way, not because I'm feeling nosy."

"Sarah? She's got no jurisdiction in France."

"I know. However, insider knowledge might come in handy."

"True. He lay at the bottom of the steps." Chris's voice was blander than a soda cracker. "First, I mistook him for a large clump of seaweed. My next thought was seaweed isn't red. It's odd, you know? How long it took to make sense of what I was seeing. Anyway, things clicked, and I dashed down the steps. Stupid of me."

"It's what any normal person would have done. I guess you checked his pulse and stuff?"

"Of course. That's what got me into trouble."

"Why?"

"Think blood on my hands."

A shiver slipped down my spine. "How did he die?"

"Must've bashed his head on the steps."

"It could've been an accident."

"That's what I thought at first. Until I spotted something shifting at the lifeguard hut. I'm dead certain there was another person around. Unfortunately, the sun was rising, and I couldn't see much with this blasted flashing and glittering going on. And, worse, that wasn't the only other insomniac in the vicinity. Some senior citizen spotted me bending over the body and had nothing better to do than to call les flics."

Tempted to point out that this too was normal civilized behavior, I bit my tongue. In his current state, Chris wouldn't take kindly to smarty-pants comments.

"Shame the good lady claims she didn't see anyone besides me."

"Hm. And you're suspect numero uno because you were there and had blood on your hands?"

"Not on my hands. I'd washed it off in the surf, but I was wearing long sleeves—mornings are rather cool here—and it stuck to my cuffs. My effort to 'hide the evidence' also counted against me. That and the fact I'd been running around for the best part of thirty minutes with no witnesses around , plus—"

"Oh, bollocks to that."

"My thoughts entirely. It gets worse. I have history with the guy, which gives me a motive."

My stomach lurched. "How?"

"Had a run-in with the moron the day before. He kept shadowing me, and it got on my tits. I confronted him in the bar where he was nursing a beer. Top that off with an appointment in the man's online calendar for a rendezvous at the beach at six—no names, of course—and voilà, you have a premeditated murder."

"How...bizarre. Both the appointment and the bit with the shadowing. Actually, why would he do this?"

"Shadow me? No idea. Still, the guy popped up absolutely everywhere. Like a paparazzo minus the telescope lens. Didn't even try to be subtle. I need my space. So, I had a go at him."

"Eh, you didn't bop him one, did you?"

Even his laugh sounded bitter now. "Do you take me for a hooligan? We swapped words. He was rather shocked by my idiomatic French, believe me."

"Ah, he was French?"

"Oui, chérie."

None of this made any sense. "What's going on here?"

"That's the million Euro question, isn't it?"

Chapter Four

Throughout the night, I was tossing and turning in my bed, Chris's words echoing in my overwrought brain. No matter what the man might say, he needed me. But the odd letters meant bad news, and as long as I didn't know what was afoot on my side of the Channel, I couldn't abandon the coven.

In the world beyond my window, dawn brightened the sky and birds were singing. Frying aromas and laughter drifting over the roof of the annex told me the guests were up, enjoying breakfast on the terrace.

I swung my legs onto the floorboards and rubbed my sanded eyes. Leave for France or stay?

Someone knocked on the door. "Myrtle? You awake?" Daisy asked.

"Almost."

My cousin slipped into the bedroom, wearing a turquoise tee that left nothing to the imagination, jeans, and a worried crease on her forehead.

My mood took a plunge into depths unseen. The last thing I

needed was Daisy wearing her drama queen crown.

Wordlessly, she waggled a white envelope. "Alma found it on the doorstep when she fetched the papers this morning. She wanted to hand it over to you. Until she spotted the name."

"There's a name on it?"

"Yes, it says D. Coldron and nothing else. No postmark. Myrtle, wake up." My cousin's voice sounded stern.

I hid a yawn behind my hand and grasped the envelope. "Sorry."

Almost weightless, the missive was stuffed to a bursting point, like mine had been. An invisible weight crushed my chest, and I could have howled with frustration. I wouldn't get to France anytime soon, not with this going on. "Did you check the contents?"

"Nope, didn't dare to. It's another of those letters, isn't it?"

"Very likely. Unless I'm mistaken, the handwriting looks like the one on Jenna's." Strange that my voice should sound so normal.

Daisy licked her lips. "You'd better open it."

Having fished my nail file from the drawer, I sliced the missive open and took a peek. No rosebuds this time, only petals, all of them wrinkled and quite dead. I jiggled the envelope, but once again there was no message. When I tapped the brittle leaves with the tip of my finger, I sensed nothing.

A short pinging noise from my nightstand interrupted the experiment. Next, my phone buzzed over the slick surface. My heart missed a beat, and I bent over to pick it up and check the caller ID.

Not Chris, but the Colonel.

"Elmsworth here," he barked. "We have a problem."

He wasn't telling me anything new. "Only one?"

"I received an envelope stuffed with rose petals. Hand-delivered like Jenna's."

"Let me put you on speaker. Daisy's also got one. With her name on it."

"Morning, Colonel." Daisy waved at the phone.

"Hello, Daisy. Got my name on it as well. Waste of time. The only other person in the cottage is Buster, and he doesn't read much."

On any another day I might have laughed. "Anyone else got a special delivery this morning?"

"Not to the best of my knowledge. I rang the Ragworts at eight a.m. sharp, and they didn't receive any letters. Started the call chain, but I bet we'll draw a blank."

"What makes you so sure?" Daisy asked.

"Of the coven members who ever hexed anything, only five are alive. That's you, Myrtle, Jenna, myself, and Rosie Ragwort, though her levitation stunt with the breakfast roll isn't common knowledge. I suspect she'll be spared."

"Looks like whoever's behind this seems to target the, eh, active witches," I said.

"Oh, wow." Pride flitted over Daisy's lovely face, smoothing the frown. How I wished she wouldn't live to regret her foray into the realm of magic.

"I'll keep you posted when I hear back from the call chain," the Colonel said. "I couldn't get hold of Emma, by the way. No one knows where she is."

My phone pinged another incoming call, its ID hidden. My heart gathered speed. "I'll ring you back, okay? Someone's trying to reach me."

"Sure. Good luck."

"Witch's Retreat B&B, good morning."

"Good morning, Ms. Coldron," Bob Ignatius said, his voice as smarmy and smooth as always.

For Daisy's benefit, I put him on speaker. "Yes?"

"I gather by now you've found out what my wayward nephew's been up to?"

I dug the nails into the soft flesh of my palms. While my gut instinct told me to cut the call, my logic insisted I should find out what he wanted. In the end, I compromised on, "Yes."

"Does he need a competent lawyer?"

Now, that wasn't what I'd expected. "He said he's got one, thanks."

Daisy rolled her eyes and drew her forefinger across her throat.

"I see. And what about the letters?"

"We're still working on those."

Static sizzled into my ear. "I bet there'll be more," Ignatius said a few seconds later. "Someone is threatening us."

As far as I was concerned, there was no "us."

"Listen, the business with my nephew in France..."

"Yes? What about it?"

Daisy cocked her fingers, pretending to fire a pistol at my phone.

"I don't like it."

Sometimes, my inner imp gets the better of me. "Isn't any publicity good publicity? Good for business, I mean."

More static. "I'm on your side, you know?"

"Mr. Ignatius, I'll believe that when you give me a reason. For the time being, we have a common problem, but it doesn't make us bosom-buddies."

"Ms. Coldron, you're too...prickly for my taste. However, we could try to be allies, hmm? At least while we're facing the same storm. If you take advice from an old foe, I'd suggest you travel to France and find out what's afoot with the Frogs."

Daisy did a great impression of fainting onto the bed.

"Why don't you go yourself?"

"I'm not exactly inconspicuous."

As much as I hated admitting it, he was right.

"Think about it. Give me a call when you're ready. Bye for now." He cut the connection.

I scowled at my phone. "I wonder..."

Daisy rolled onto her belly and dangled her legs in the air. "What is it?"

"Those letters. Especially the ones posted in Carnac. I

thought I mustn't leave because of them, but what if someone is trying to sow confusion and keep me away from Chris?"

"That would be a rather wacky way of going about things." As if jolted by an electric current, she rolled over and sat. "Snap. I must be getting old. Sarah's downstairs. She wanted to have a word with you."

"Eek, how long has she been around? I haven't showered."

Daisy checked her watch. "Not long. She's having the full English, so take your time."

When did I last raise Alma and Cecily's salaries? Two months ago? I'd have to do it again. "Hide the envelope somewhere safe, please. I'll be down in a jiffy."

In the bathroom, I grabbed the nearest bottle of shower gel—and ended up covered in gummy bear-scented foam. What the heck made me buy that? Once I was dry again, I slipped into my undies and sprinted to Auntie's wicker wardrobe to throw on red capris, a white tee, and a pair of blue striped espadrilles. Not the most brilliant of color combinations, since it reminded me of the trouble brewing in France.

My strawberry hair fluffed in all directions, so I tied it back with a scarf. Putting on my face in my current mood was out of the question—I'd end up as a London Dungeon exhibit. Instead, I charged down the stairs au naturel and found Sarah in the conservatory, close to the open French doors.

"Ah, I was wondering where you were. Have a seat." She let her gaze travel over me. "What's it with the pirate look?"

She, of course, appeared as smart as any Parisian in her grape-red pantsuit and short-sleeved cream blouse.

I sat and adjusted my headscarf. "My personal dresser is having a day off."

"More like a week." She grinned.

Aromas of bacon and coffee filled the room, and my stomach growled.

Sarah pushed the breadbasket across. "Help yourself; there's plenty. Sadly, I don't have much to report. The gendarme I talked to became super helpful once I claimed Chris as an acquaintance. Wanted to know if I had any dirt on him."

"Innocent until proven guilty, much?"

"Fret not, I acted the dumbo. It looks as your Chris took a massive dislike to the victim, one Eric Poussin, private investigator by trade, and that's what made him the key suspect."

"If the man is...was a private snoop, this might explain why he was shadowing Chris, who was mega pissed off by the paparazzi stunt, for sure."

Sarah looked up. "You heard from your man?"

"Early this morning."

"Great. Begs the question of why he was being followed."

As if on cue, the kitchen door banged open and Cecily entered, bearing a teapot and a wide smile. "Here's your tea. Would you like anything else?"

"Not now. Thank you anyway."

"Always a treasure," she said and waltzed off.

Once she had left, Sarah leaned in. "The real problem seems to be the argument on the beach."

"What argument?"

"Well, the same senior citizen who called the gendarmerie said she spotted two people at the top of the steps earlier. It looked like they were quarreling. She didn't have her glasses on, and the sun wasn't up yet, so she had difficulties working out the details. But she claims she saw two people, one taller than the other."

"How could she know she was watching an argument when she had vision issues?"

"Their body language came across as aggressive. Madame left to fetch her glasses and her phone, which took a while, and only then did she spot Chris, bending over the man on the ground. When asked if he was the same person she saw before, she wasn't sure at first, but eventually she claimed she was, and she's stuck to that ever since."

Fear knifed my guts. This wasn't looking good.

"Chris thinks he spotted a third person."

"Yes, he told them. Unfortunately, he doesn't have a witness. I'm sorry to say, but if I were the investigating officer, I'd home in on your man as my prime suspect. They didn't say so, but it was implied."

The roll tasted of ash, but I forced myself to finish it. Too many shocks on an empty stomach would result in indigestion, and I needed my wits about me.

Sarah ran her fingers through her spiky dark hair. "Any mitigating circumstances?"

The answer to that was no. Unless… No, I couldn't share the intel concerning the envelopes. Only the direst of emergencies would make me expose the coven. Chris might need help, but I wasn't desperate yet.

"He's no killer."

"You know that, and I agree. My French colleagues released your man because everything's a bit shaky right now, but if they find more evidence against him, he'll land in the slammer. Not sure if they'll let him post bail either. If I was a magistrate, I wouldn't." She cleared her throat. "As a police officer, I shouldn't be saying this, but it might help if you headed for France and embarked on some sleuthing."

Put like that, what choice did I have?

—

Sarah's departure was rather abrupt—duty called, and she had to leave for the Swindon outlet center, where some moron was causing an upset in the lingerie shop. I picked up the plates and carried them into the kitchen.

Wrong move.

"They say Mr. Lentulus has been arrested?" Alma had her voice of doom routine down pat.

Seriously, life in a small village was not for the faint-hearted. I channeled the late Queen and slipped her my haughtiest look. "Are they now snooping into my private affairs?"

She sniggered. "No, Anna speaks French and watches their news."

Anna, our churchwarden, wasn't exactly running a Myrtle fan club since I'd caught her sabotaging my guests. A dull throb started in my head. "It's a misunderstanding. But I might have to go to France. To help him."

"Of course, you do." Cecily said. "Never mind us. There's not a lot going on this week."

Alma studied the calendar stuck to the commercial fridge with a magnet shaped like a witch's hat. "And the next. Mostly salespeople and a few hikers. Oh, and the two Druids are back. They're kind people."

"Ack, I can't imagine it would take this long."

Cecily's mien turned to woeful. She wrung her rubber-gloved hands. "You never know where you are in these foreign places."

"They have maps for this," Alma said.

"That's not what I meant," Cecily responded.

The pounding in my head increased the beat. "Thank you, ladies. I appreciate the support. Now, I've got some planning to do."

My artificial smile slipped the moment I entered the corridor. It collapsed into a smoking ruin once I'd shut the living room door behind me.

"Can someone cut me a break, please?" I said to no one in particular.

As was her want, Tiddles ignored me and kept washing herself, her pink tongue doing overtime on the tortoiseshell fur. She then stretched out a leg and nibbled the space between her overlong claws. Since any attempt at cutting them met with me getting scratched and her being thoroughly miffed, I'd left the claws well alone ever since.

Petty, bless her blossomed soul, however, rose on a sparkly

vortex and lifted her pot up to the level of my face.

"I need your advice."

My familiar fired a questioning spark.

"Chris is in real trouble, and no matter what he says, he needs me in France."

Petty leveled her pot on the table and knocked once for yes.

"Yeah, that's what I thought. But these letters... See, Daisy and Elmsworth received some as well. Hand-delivered, not sent from Carnac. Sure, I know my friends are competent people, but leaving them in the lurch feels wrong, especially as this might be exactly the desired effect. What if someone's engineering trouble to lure me away? On the other hand, the purpose of the exercise might be the exact opposite, and someone wants to stop me from helping Chris."

Sparks ran along Petty's leaves and winked out.

"Heavens, no matter what I do, it's bound to be wrong. Oh, Petty, what am I supposed to do?" My voice wobbled, and the blasted tears were back.

Petty fired off her rose fragrance that symbolized my union with Chris. Then she hovered across and nudged the jeans pocket containing my smartphone.

"No, calling is not enough. I need to be with him."

Petty's knocked the table once, her "yes" cracking into the room like a shot. She wriggled her flowers, and now her lemony, happy scent flooded my nostrils.

"You're quite determined I should go, are you?"

Another sharp rap rang out. Once more, Petty nudged the pocket containing my phone.

"Who am I supposed to ring...oh, you're telling me I can easily stay in touch with the others?"

One rapped "yes," followed by a flurry of rustling and whooshing noises, was her reply. The pot bounced across to the table and pirouetted around.

Tiddles opened one disapproving eye.

"Don't worry, the Simpkins sisters will feed you. Petty's

right. To heck with the letters. I must go. Daisy'll care for Petty—"

The pot shot up, blooms jiggling. My familiar went invisible, winked back into sight, and shot through the room. Two raps vibrated from the table.

"No? But you like Daisy?"

Rap. My flowery pal spun dizzying circles around me.

"Are you saying you want to come too?"

Yes, she rapped.

"Hmm. Maybe that's not such a bad idea."

The lemon scent increased to the point of unpleasantness. I sneezed, which loosened a few blockages. "You want me to take Daisy and you?"

Petty rapped out another yes and spun around in a circle, sparks flying in all directions.

Oh, what the heck, I needed whatever support I could get.

Chapter Five

"**O**h, Myr, look. A baguette-o-mat." Daisy pointed at something beside the road, her arm blocking my view. The guidebook she'd been studying tumbled into the footwell.

"How am I supposed to drive when your arm's in the way?"

"Oopsie." She withdrew the offending limb.

Petty, now inhabiting a chartreuse plastic bowl—it took up less space than her fake terracotta pot—reflected in my van's rearview mirror.

"Down, Petty, for heaven's sake. There are cars behind us. What are people supposed to think if they see you? That we're rejects from the Chelsea flower show?"

My familiar vanished from sight. I wiped my sweating brow.

After twenty-four frantic hours filled with travel arrangements, coven and housekeeper liaisons, and sweet-talking Chris into accepting my arrival on his French doorstep, we'd made it to Cherbourg without any major mishaps. Minor glitches, such as the heavy traffic, driving on the wrong side of the road, or Petty and Daisy between them filling the van's interior with clashing fragrances, didn't count. Getting Petty to agree to my code of conduct for sentient houseplants on vacation was a lot more difficult. She wanted to leave the van—invisible, of course—while I wanted her to stay in the vehicle. I got my way in the end, but my magical flower was acting as jittery as an incontinent canine. Seriously, traveling with a dog was a cushy number compared to the challenge of keeping my nosy and clever zombie plant under wraps.

In a way, I'd known schlepping Petty along wasn't exactly a brilliant idea, but without the little menace I felt sort of incomplete. "Um, what's a baguette-o-mat when it's at home?"

"Big display cabinet with baguettes in them. You slot in a Euro or two and out comes your bread."

"Seriously? Hmm, this is France, after all. I guess running out of baguette is a no-no."

Daisy now pointed at an object whooshing past on the right. "A standing stone. Myrtle, that was a standing stone." Her voice was too high-pitched and too loud for my small van.

"Less of the decibels, if you please. If you squeal every time you see a rock, it'll get a bit noisy in here. Carnac is famous for its stones."

"The alignments," Daisy said. "They're called that because they're arranged in long rows."

"Yep, Chris and I wanted to visit those. Like the rest, they'll have to wait."

At least I was on my way, rushing down a road marbled with the shadows of beech leaves, the blue of the skies enhanced by puffy, white clouds and dotted with seagulls. As my van rumbled away the miles, a fresh countryside scent wafted in through the passenger window. Soon, I would see Chris. No matter what the Fates threw at us, we'd be together.

That was assuming the police didn't arrest him first.

Daisy patted my thigh. "Oh, you'll sort this mess out in no time, and there'll be plenty of opportunities for sightseeing. There's also dolmen and single stones here, called menhirs.

Like in Asterix and Obelix."

I tossed my cousin an amused glance. Her cheeks flushed, wisps of auburn hair having slipped from her braid and caressing her cheeks, she looked like a teenager, not a woman in her twenties.

She twisted around in her seat and gave me puppy eyes. "Do you think this place also has witches?"

"If it does, they must be experts in hiding. Jochen's researchers tried absolutely every trick in the book, but they never flushed out any more magical beings. Not living ones, anyway."

"There's one. Look, Myr. Over there. Can't you see?"

I spotted plenty of trees and a sunlit road gently dipping down a hill. "What? Where? A witch?"

"No, silly. A dolmen."

"Daisy, you're shifting topics so fast, it's too much for little old me."

She burst into laughter. "You're only thirty, Myr. That's not too bad."

"Thanks, I guess."

A white town sign with black lettering popped into view.

"We're here," Daisy said, clapping her hands. From behind, a pink spark sailed over my shoulder.

"Cool it, girls, this is Carnac Le Bourg. We want Carnac Beach. Yikes." I stomped on the brakes. People were milling over the road, carrying baskets filled with bottles, vegetables, jars, packets, and parcels. To my right, colorful sun umbrellas shading stands that offered everything from cheese to beach towels indicated we'd arrived on market day.

"Oooh. They've got perfumes and soaps. And clothes. And jewelry. Oh, and look at this sweet stand."

"I refuse to look at sweet stands. Daise, we're not here for the shopping."

"Bah." Daisy was busy fiddling with her smartphone. "Yay, they'll be open on Sunday again. By that time, you'll have flushed out the killer, and we can have a great time."

Her faith in me was touching. It was also sort of scary.

The last miles crawled along. Carnac was a thirty zone—kilometers not miles—and it took half a lifetime to reach its lower part. Without the navigation app, I would never have found my way through this maze of streets crisscrossing each other. While I couldn't see the ocean, I could certainly smell it; the tangy whiff of seaweed and ozone was unmistakable. As if to tease me, the app guided us up the hill again. If I'd been in charge of choosing a hotel, I would have gone for one right next to the sea. Chris apparently had other priorities.

Chris, soon you'll see Chris. The van gathered speed, so I forced myself to take my foot off the gas.

"There," Daisy squealed. "Auberge du Dolmen one kilometer."

Once more, the van sped up, prodded by my impatient foot. To hell with speed limits.

Soon after, I steered the van into a paved car park. Behind it waited a square building, white with a dark roof like the local stone houses, but without their charm. The architecture screamed seventies, dished up with a heavy dose of concrete. The place was well-maintained, the red geraniums lining the balconies adding a nice splash of color, but it wasn't up to par with my vision of roses creeping up white walls, gauzy curtains fluttering in the wind, and four-poster beds. I couldn't stop the sigh from rushing past my teeth.

"Yuck. Not my taste, to be honest." Daisy scrambled from her seat and stretched.

"Mh, yes. Bit—mundane. But it's where Chris is."

Yay, and here came the man himself, bursting from a revolving door and chasing down the steps leading to reception.

"Chris!" I catapulted from the car and threw myself into his welcoming arms. They wrapped around me, cocooning me in his spicy warmth.

Until he kissed my ear, sighed, and took a step back. "Never thought I could miss anyone so much. But is this wise?"

"No. Yes. Who cares? We're here."

A corner of his mouth kicked up. "You are indeed. Hi Daisy. Right, let's get you sorted, shall we?" He glared at our luggage. "Are you planning to stay until doomsday?"

My heart skipped and hopped. Never had Chris's sarcastic comments been more welcome. "Only as long as it takes."

He smiled. "I love you, you know?" He grabbed two suitcases, one of mine and one of Daisy's, and raced back up the steps.

—

His arms crossed behind his head, Chris lay on the double bed, watching my progress with a bemused expression on his stubbly face. "You forgot the kitchen sink."

"Didn't need it. This, however, might come in handy." I lifted the recipe book from a cedar box lined with plushy red velvet, which once harbored two bottles of aged port, a present of a generous and very happy sales rep. "Guess who else is here—"

Accompanied by a firework of sparks and a heady rose scent, my familiar exploded into view.

Chris snapped from his slouch. "Blimey, why did you drag her along? Your cousin I can sort of understand. Moral support is important, as is a second driver. But your flower?"

Petty's leaves drooped.

Chris groaned and buried his face in his hands. However, he spoiled the effect by peeking through his fingers. "Sorry, sweetheart. No offense meant. But isn't this a tad risky? What if someone spots you? Like the chambermaid, for example."

Her leaves springy once more, my primula hovered over to the bed. Chris laughed and flicked his fingers at a pink blossom. "Missed you, little one. You and your mistress."

"Especially the mistress, I should hope." I dropped onto the bedspread, doing my best to ignore the psychedelic zigzag pattern in white and black. The entire suite was supremely monochromatic. Black and smoked-glass furniture against whitewashed walls, the only relief being the lime-green curtains. The place couldn't be more different from my gauzy visions if it tried.

Stop moping. This isn't a holiday. My inner voice sounded a lot like my late aunt.

Chris slumped into the pillows and stared at the ceiling. "As happy as I might be about your arrival, I'm not sure how long I'll be able to pester you with my grumpy self."

My heart fisted. "Where do you think you're going?"

He rolled over, his expression bleak. "Nowhere. Not allowed to leave the area. By the way, I'm referring to my arrest. They're closing in on me, Myr. This is for real."

A sharp pain sliced into my stomach. I'd known. This was my reason for coming here. And yet his comment, voiced in a bland tone, cut deep.

"Did the solicitor get in touch?"

"Yes, Maître Kerluac is a professional, tough as boots and convinced I did it, no matter what I tell him. My wretched temper speaks against me, I'm afraid."

My mouth suddenly as dry as cat kibble, I said, "By that you mean the witness, correct? And the fact she observed you having this argument with the dead guy?"

Chris slammed his fist on the bedspread. "She saw nothing. If she did, it wasn't me."

"Witnesses are notoriously unreliable. Surely, the cops know this."

"She claimed it was me she saw with the victim. Twice. The cops, of course, lapped it up."

"Why would you kill the guy? Only because he annoyed you with his snooping? I don't buy it."

Chris shrugged. "They're wavering between manslaughter and murder. The inspector is eager to make it the latter—that I arranged to meet Monsieur Putain with the intention to finish him off."

"Poussin."

"Play on words, dear."

"Wasted on me. My French is wonky. German's my preferred lingo."

"Ein Bier, bitte is all you'll ever hear me say. Anyway, according to Monsieur Flic, I'm guilty of shoving the guy, preferably with evil intent."

"Come on. They're grasping at straws."

"They're good at it, believe me."

There was no point in making soothing remarks. In his current state, he wouldn't listen. A distraction was needed, so I filled him in on the rose petal mystery, carefully talking my way around his blasted uncle and finishing with, "I don't have an inkling why this is happening. However, it bothers me—a lot—that my letter and the one sent to your uncle were mailed from Carnac." I bit my tongue, but it was too late.

Chris sat bolt upright. "Did you say uncle? As in my Uncle Bob?"

Open mouth, insert foot. "Yes. Only in his case, the envelope contained rose ash, not petals."

Chris's dark brows slanted, giving him a distinct satyr look. "He might be at the origin of all this hassle."

"He might, yes, though I'm pretty sure he isn't. Under his slick operator facade, he was upset. That's why he visited in person. And for once he didn't suggest I should hex for him. Oh, and he was the first to tell me about your murder problem."

Chris scratched his bristly cheek. "Bizarre, but quite honestly, I don't have the energy to worry about Uncle Bob on top of everything else. Okay, Sherlock, where would you like to start? Oh, what about your cousin?"

"Daisy will be there for us if we need her, she said. In the meantime, she'll go shopping."

"That's kind of her."

Warmth flooded my chest. Yes, it was. "Let's start with that beach where...well, you know."

My primula rose on a pillar of bright light.

"Oh, no, you can't come. We discussed this before we left, remember?"

"Surely, she'll be okay on the balcony." Chris yanked the heavy door open, making the curtains flap into the room. "As long as she stays invisible. Can you keep it up long enough, sweets?"

Yes, Petty rapped. Then she snapped out of view with only the billowing curtain to tell where she'd gone.

I rose from the sofa—and hesitated. "Actually, should you be out there?"

Chris quirked a brow. "I'm not under house arrest. Also, the crime scene has been cleared. Now's our best chance."

—

Over trails springy with pine needles we trotted, passing rough stone walls teeming with blue and pink hydrangea and the slate-covered cottages sporting chimneys at each end that are so typical for Brittany. However, when the shadow of an airplane slipped over the path, the rumble of distant engines morphed into an ominous growl. As lovely as the place might be, what should have been our first holiday together was jinxed.

Chris's arm, heavy and warm, encircled my waist, but the frown line on his forehead and his pinched face told me his mind was miles away. With a murder accusation dangling over him like the proverbial sword, who would blame him? Not me, for sure, so I left him to his thoughts.

When we hit upon a crossing where a sign promised "Beach, 500 meters," he stopped and looked at me. "I'm sorry."

"What for?"

"This isn't what I planned for us. It's why I didn't want you to come. Now the place is spoiled for you."

While I might have been thinking just that, I wouldn't tell him. "You can't seriously expect me to continue with my life as if nothing happened while you're fighting off law enforcement. I'm here. It's what it is. We'll sort it out, even if it takes every spark of my blasted magic."

Chris laughed, drawing me against his chest. His heart pulsed in a reassuring, steady rhythm, and we lost ourselves in the fragrant warmth of a summer day until applause pattered in from a passing car, accompanied by loud hooting.

The spell broken, I took a step back and searched his still-smiling face. "Vive la France!"

This time, he laughed outright. "Gods, how I missed you, Myr."

A lot more relaxed, we continued our trip until the beach shifted into view. Under different circumstances, the half-moon bay opening before us, featuring cliffs topped by colorful cottages on one side and a low-slung peninsula on the other, would have made my soul whoop with happiness. Today, however, the sunlit scene was tainted by the knowledge of a life cut short, one in a string of strange incidents which threw long shadows over my life.

The tide was out. A flotilla of boats lay stranded on their sides, watery runnels underneath snaking over the sand like radiant veins. Seagulls screeched, dipping and diving at the tidal pools and the unfortunate critters trapped within.

"Voilà, welcome to Plage du Beaumer," Chris said.

"Lovely place. Well, it would be, if it weren't for the murder." My roving gaze found a weathered wooden structure covered in peeling white paint. One side bore a red cross. I pointed. "What's this?"

"Lifeguard hut. It's where I noticed another person, only to unsee them the next moment because the blasted sun was blinding me."

"Mh." As if pulled by an elastic band, my legs crunched over the rough sand, headed for the hut. Arrived at the rickety building, I shaded my eyes to take in my surroundings. "Where are those stairs?"

Chris tromped up. Wordlessly, he pointed to the left of the shed, where a rusted handrail accompanied broken concrete steps down to the beach. The moment I turned my head, a jogger sprinted up from the sand and sped uphill.

Now, that was interesting. "You can see both the beach and the steps from the hut. And those houses on the cliff. Makes me wonder whether someone was watching you while you were trying to help the poor guy." I leaned against the weathered boards of the hut, warmed by the sun—

Unfair.

From nowhere, the word whispered through my mind, quiet like an indrawn breath. But the thought wasn't mine. Nor was the urge to hurt, mangle, and destroy.

My heartbeat spiked. Bruise-colored spots blurred my vision, and I staggered aside.

What the—

Unfair. Hate you. Serves you right. Glee puffed up my chest.

Sucked into a roaring noise, the alien thoughts and sensations cut off. Scents of fresh soil, wet leaves, and spring flowers overpowered my nose, while a bright, green light rolled in the same moment a silky object brushed over the palms of my hand.

"Whoa, steady old girl," Chris said from far away.

I blinked. The greenness vanished. The roaring segued into the sound of the ocean, a glittering expanse of blue far, far beyond the bay. A stiff breeze tossed my hair, and the stink of rotting seaweed elbowed aside the scent that accompanied my skylles.

Pale pink petals materialized all around me, only to be carried away on the wind. As I watched, the flurry slowed. When it stopped, the strange thoughts and sensations had vanished.

"Eep, hexing attack." My voice sounded gravelly, as if it belonged to someone much older. I cleared my throat. "Sorry."

"What brought that on?"

"I...I'm not sure. I sensed...someone hating me? Someone feeling smug about something?"

His brows slanted. "Now you sound like your cousin. You sensed hatred and smugness, or you didn't?"

"Well, I sort of did. Someone hates me because something's unfair, perhaps? Yes, I know I sound like Daisy. Consider me baffled. I don't understand in the slightest what happened just now. I heard words and experienced emotions which weren't my own. Once my skylles rose, the sensations stopped. Just like that."

Chris stared at the last petals, some slipping over the sand, some snagging on the rough boards of the hut. "Sounds like a new trick. Luckily, no one was around to watch. But—"

"It's not as if I planned this."

"You're a witch. You're supposed to have extrasensory perception, right? The experience might've been triggered by your unruly skylles."

"Usually, they rise for a reason."

"Ask Petty. She seems to understand magic better than you do."

"Har, har."

The banter felt familiar, normal. However, normality had gone out with the tide. Not only was this place a crime scene, it also seemed to be a paranormal hot spot. Was it too much to hope the two wouldn't be related?

Chris drilled the tip of his sneaker into the sand. "If you picked up someone's emotions, it means I wasn't hallucinating. I did see someone."

"Yup. That's assuming the sensations I experienced are connected to the person you noticed."

A thoughtful expression sneaked onto Chris's face. "Also, this would mean emotions can be left behind for you to pick up. How would that work?"

"Search me. As you said, Petty might know. I, for sure, haven't got the foggiest."

"Can you try again?"

Should I? If it helped Chris, I'd better have a go. "Hang on." I willed my muscles to relax and unclenched my fingers, bracing myself for impact.

Skylles, can you call up these odd thoughts and sensations?

Seagulls wheeled, children laughed and splashed in the shallows. An ice cream vendor hollered his way along the beach.

Otherwise, nothing happened.

"Nope, didn't work."

I dropped my hands—and a petal slipped out, see-sawing to the ground. Under my feet, the ground vibrated, the sands shifted and churned, making me lose my balance. The greenness was back, fizzing from one ear to another, and whatever had been up now seemed to be down. When my flailing hands scraped over the rough wood of the shed, my world once again settled on its axis.

"Ow." I examined my stinging palms for splinters but found none.

Serves you right. Hate you.

Like a chilled spiderweb, the words blew through my mind, and with it came a sensation of wrongness. Someone unclean had been where we stood now. Someone who soiled the weathered wood with their hateful thoughts. Did I touch it before the magical outburst? I must have, and that triggered my skylles? Because that person disliked me with a passion?

No, that made no sense whatsoever. I hadn't been around when Chris spotted the person most likely responsible for those nasty thoughts and emotions. Instead, the remembered spitefulness must have been directed at the victim.

A chill skittered down my spine.

Or Chris, for that matter.

As usual, I had no proof. But there was a lot of sense in this line of thinking, outlandish as it might seem.

Hunches, that's me. I rubbed my hands on my shorts, desperate to rid myself of a slime existing only in my imagination. I would have to get to the bottom of this mystery; otherwise, I'd never find peace.

"You were probably right, you know?" I said.

"Always. About what?"

"You spotted someone by the shed. And I fear this person might well be our killer."

Chapter Six

C hris shoved his hands into the pockets of his black shorts and stepped away from the lifeguard hut. Apart from a few stray petals caught in tufts of pale grass, there was no visible trace of my brush-up with magic. The building behind me, its boards bleached and peeling, was just that—a building.

Yet, I'd learned to trust my skylles. As wobbly as they might be, if they flared up, it was for a reason. While I didn't understand what was going on here, something certainly was. I'd have to rely on both my reason and my magic and take it from there.

I meandered after Chris, picking up an ice cream wrapper half-buried in the sand. Litterbugs everywhere. For want of a bin, I wadded the paper and shoved it into my pocket.

"Shame your special gift won't count as evidence," Chris said over his shoulder.

"No, you're right. We need facts."

Chris slapped his forehead. "Oh, Lentulus, you're an idiot."

"No, you're not."

"In this case, yes. Should've thought of this a lot earlier."

"Care to enlighten me?"

He raised his arm with the new watch and tapped its shining surface. "This little baby."

I wasn't getting it, and my face must have spoken volumes.

"I told you the thing is hot off the factory, all-singing, all-dancing."

"Doesn't look cheerful to me. It's black, like most of your clothes. So?"

Chris grimaced and tapped the surface again. "It's got multiple functionalities, with photos being one of them."

Finally, understanding dawned. "You have a photo of the person you saw? Please say yes."

He grinned. "I might. See, I took photos of the body, complete with a time stamp."

"Uh oh, why?"

"At the time I didn't realize I was being watched, but I had a nasty feeling I might be in trouble. No need to tell me. I know it doesn't make much sense. I guess I wasn't thinking clearly. Anyway, I flipped through the images once and gave up quickly. They're—not nice. I noticed one thing, though."

"Like what?"

"It appears I didn't switch the photo function off. The watch does that by itself after a minute, otherwise, even the cloud storage would run out of space. This model is cool, ensures auto-uploads every five seconds while on and—"

The techie-babble vaulted straight over my head, but there'd been some useful nuggets among his spiel. "Cool down, you're losing me. Basically, you're telling me the gimmick was active when you spotted this person, and you might have a photo but didn't check for it yet, correct?"

He grinned. "Yes. What's even better—les flics weren't fazed about the watch. Never bothered to with it."

I craned my neck to get a closer look at the gizmo. At first sight, there was nothing spectacular about it. Elastic wristband made of some matte-black material, shiny black screen twice the size of a postage stamp, displaying a series of icons in red, white, and blue plus numerals.

"Chris, we should check the photos. Like now. Not the ones of the private dick, obviously, but the ones with the hut. If there are any."

Confusion crept onto his face. "Which private dick would that be?"

"Poussin. The guy snooping after you. Oh, didn't I mention that? Sarah told me."

"No, you didn't share your insights. She thinks he might've been paid for his troubles?"

"Looks like it. Which begs the question of how a dead PI, a mysterious person, and your presence at the beach link together? And who's behind it?"

Chris drummed his fingers on his watch. "Then we better return to the auberge and search the pics. We need higher resolution for that."

A bunch of kids in swimming trunks raced past us, kicking up sand and whooping at the top of their voices.

"Doesn't your watch's display turn into a portable fifty-five-inch screen at the push of a button? I'm disappointed."

Chris half-smiled. "You wish. Here, I'll show you." He fingered his phone from his pocket and swished his way through the photos. "Look, there's the hut. Bit fuzzy, must have been moving my arm. Can you make out any details? I can't."

Nor could I. "You're right. We'll need your computer."

"Talking of computer. If the cops confiscate it, we're in trouble. Oh, blast." He paced on the sand. Then he stopped. "Myr, can you memorize something for me?"

"I can try. What is it?"

"Blue Poodle Marmalade Pettyzombiemagic. All words with caps, but Pettyzombiemagic is one word."

"What?"

The grin transformed his face. "Blue Poodle Marmalade

Pettyzombiemagic. It's my primary password for the non-business stuff. The other ones are more complicated."

"On your laptop?"

He shook his head. "Everything's in the cloud. May I have your clunker...pardon, your phone, please? I'll enter a link, which will show in your search history afterward."

While he was tapping around on my phone, a warm glow spread in my chest. The super hacker was entrusting me with the code to his private files. How cool was that? Who needed roses when they could be given passwords? Love in the twenty-first century was certainly unique. "Shouldn't a password have plenty of numbers and freaky characters?"

"Makes them hard to remember and requires password safes. Which you also need to protect. I'm not a fan of them. A random sequence of unrelated words is quite hard to crack and much easier to remember."

He was right. As I muttered the words under my breath, they wormed straight into my brain. "Uh, why the honor?"

"I fear my arrest is only a question of time. In that case, I'd be surprised if they didn't confiscate my hardware. This way, you can check the camera files yourself on your tablet. Its screen should just about do the trick."

How could he talk of an arrest? I'd come all the way from the UK to stop this from happening. But then, I couldn't work miracles, could I? Infusing fake cheer into my voice, I said, "Okay. I hope I won't delete anything."

He rolled his eyes. "Don't, I beg you. Make sure your cousin's around. She seems to be quite clued up with technology."

"Your wish is my command, though it wouldn't hurt to tell your lawyer about those pics."

"I will. Anything that screws with law enforcement's idiotic theories is good. Do you need to see more, or can we leave? This place gives me bad vibes."

While I agreed, we weren't quite done yet. "Let me have a

quick sniff of the steps. Maybe it triggers another insight."

"Only if you promise not to have another hexing fit and spray your petals all over the place."

Marching ahead, I said over my shoulder, "Then you tell them we're practicing for the wedding."

Something along the lines of "last thing I need" drifted after me. He might have spared himself the breath, for the broken concrete steps bearing witness to the end of a life gave me— nothing.

I ran my hands over the railing, crusted with rust and sea salt. No whispers, no anger. No skylles rising. Nothing.

"And?" Chris asked.

"Nope. Sixth sense is on the blink. Whatever triggers me is somehow related to the hut. Which is a bit weird, since the PI died at these steps. One would've assumed...Oh, well, worth a try, I guess."

Chris shrugged. "Search me. Would you mind if we moved on?"

A cool breeze sneaked under my T-shirt, and I shuddered. The place was spooky beyond words. "Sure. Can't wait to have a go at those photos."

"If you don't mind, I'd like to wander around a little longer. One never knows... And you haven't seen the big beach yet. It's quite something."

I translated that into, "I want to enjoy the fresh air while I can." With a heavy heart, I interlinked my arm with his. "Love it."

The photos would have to wait.

—

Carnac's main beach formed a golden crescent, framed by a row of mansions, some of which appeared Victorian in design with their pointy turrets and wrought-iron balconies. The whole setup reminded me of an old-fashioned British seaside resort, minus the pier and interspersed with jarring examples of seventies architecture. To reach the water, we had to walk barefoot over the rippling mudflats, our shoes safe in Chris's backpack. Arriving at the surf line, we dipped our toes into waters too glacial to contemplate a swim. It didn't matter; we were here; we were together, and Chris was almost back to his usual logical self. Almost—his mood marker was still quivering on the black line.

"Maybe I should ring Sarah," I suggested.

"What for? She can't do anything. I'm amazed these connards shared any sort of information with an outsider."

"That's because she's also a copper. What's a connard?"

"Forget the word. It's rude." Chris slipped on a sludgy bit and swore under his breath. Whatever unsuitable comments he was making were swallowed by the phone trilling in my pocket. I fished it out and checked the caller ID.

"Hah, guess who's calling. Must have been my witchy vibes." I thumbed the call open. "Hello, Sarah."

"Myrtle? Hello? Reception's a bit wonky."

"I'm on the beach. With Chris."

"Ah. Right. I hope he's not listening in."

Cold and hard, fear lodged in my throat. "Why? What's up?"

"The local inspector rang again. He's found out your Chris was a suspect once in your aunt's case."

"Hello? He wasn't the killer."

"The simple fact your man came under suspicion was quite enough. The inspector is digging for dirt, and he's damn good at it."

I looked up, but Chris had ambled off and was picking up seashells he then flung into the wavelets. He was an intelligent man; even with hearing only one side of the conversation, he would have guessed where this was headed.

"All this on the evidence of a half-blind woman?"

Chris's broad back tensed. He must have the hearing of a bat.

"Madame Guillou's not half-blind. She needs her specs, and

if I need to read much more of this blasted paperwork, so will I. By the way, you didn't get her name from me."

"What name?"

"Attagirl. Anyway, she might be a witness, but she wasn't exactly close to the scene. Believe me, these guys aren't crazy. They knew from the start the evidence is somewhat shaky, the reason your Chris is still free. However, what they found now is going to shift the balance."

The freezing ball in my throat swelled. "What is it?"

"Mister Poussin had a file on Mister Lentulus. He seems to have quite a few French customers, correct?"

"Uh, I wouldn't know. He's certainly fluent in the language."

"Take it as a given. Poussin was suspecting your man of illegal IT operations. He's supposed to have hacked his way into French businesses."

"He does this for a living. With the consent of the owners, mind you."

"That's where facts look a bit fuzzy around the edges. Anyway, there was a folder containing paper copies of emails and handwritten notes that indicated Poussin had dirt on your Chris. Given the argument, he appears to have known and taken offense." Sarah might sound neutral, but there was an irritated undertone in her voice that told me she didn't buy the evidence.

"What do you think?"

"Not being in charge, I shouldn't be thinking anything, but I wonder whether the folder might have been planted. Its contents seem to be nicely vague, wonderfully finger-pointing, and totally insubstantial. Looks like some serious mud-flinging's going on in the hope something sticks."

"I'll warn him, thanks. Oh, we discovered some bits you don't know yet." I filled her in on the pictures we still had to sift through.

"Hm," she said once I'd finished. "That's interesting. If—big if—you can prove there was another person in the vicinity, then my esteemed French colleagues should be investigating that lead."

"Should being the operative word. Do I assume correctly that certain officers are less likely than others to let go of their chief suspect?"

Sarah growled. "Those things happen even if they shouldn't. And I fear you're dealing with such a person, which means Chris is not on terra firma yet. Not by a long way."

I eyed the wet sand squelching between my toes. "Looks like I'd better crank up my sleuthing."

"Unfortunately, yes. Do your damnedest to prove Chris wasn't the only person who could have harmed the vic. Make sure to beam the intel to your lawyer. It's your best bet. My French colleague was ranting and raving about him, which tells me the guy knows his job and will ensure the cops do follow other lines of inquiry."

"That's what Chris is paying him for."

"Still, I reckon it won't spare him a stunt in the slammer. He's a foreigner, so there's a certain flight risk."

"He's not going to run."

"You know that. He knows that, and I believe you. But if I were local law enforcement, I wouldn't run any risks. Have a go at those pictures, and get the evidence to your lawyer ASAP. Anyway, you don't know any of this. Certainly not from me."

"Understood."

"Oh, one more thing. The good inspector still must get his arrest warrant. Which gives you a short period of grace."

"How short?"

"Not sure, but I'd say you're talking hours rather than days. I suggest you use that to brush up your evidence. Sorry, but there isn't more I can do for you now."

While it wasn't halfway enough, she'd already done a lot more than she should. If only Sarah wasn't bound by her rules and regs. If only she could take over the investigation—

Like one of Petty's sparkles, an idea popped into my mind. "I wish this were a murder mystery."

"I beg your pardon?"

"If it were, I could visit the office of that detective and comb his office for stuff those Keystone Cops might have overlooked." Like some clues about the person possibly masterminding this crap.

"Myrtle, don't even think of going there," Sarah's voice rang with alarm. "Even if the inspector's a nuisance, he strikes me as being reasonably competent. He wouldn't have forgotten anything. You'd get yourself into serious trouble."

"Don't worry. Just a silly thought that flew into my mind." Stupid it might be, but after landing it had already seeded and was growing a whole root system.

"Yeah, it's silly all right. Don't dream of pulling such a stunt. Anyway, that's it for the moment."

"The warning is appreciated. Without you, the French cops would've caught us unprepared."

Chris had returned, and upon hearing my comment, he blanched. I finished the call and dredged up a false cheer. "Right, looks like the situation's going to get worse before it gets better."

"What's that supposed to mean?"

I sucked in a deep breath and shared Sarah's intel. Storm clouds crowded Chris's forehead, and there was a dangerous glint in his eyes. He squeezed his lips shut, but the way he balled his fists spoke volumes. Somehow, his icy calm scared me more than any outburst would have done.

"Just as I guessed," he said.

My heart went out to him. "Guessing and knowing are two different things. We better shift our posteriors back to the hotel and go through the pictures. Sarah thinks the cops are due to arrive anytime."

"Do it when I'm gone. Right now, I can't concentrate. I better pack a case instead. And I want a decent dinner. This might be France, but I doubt the kitchen in the slammer is going to be Michelin standard." He braved a smile, but it dissolved into fragments.

This was all so *wrong*. As much as I would have loved to vent my frustration, screaming wouldn't make things right. At least one of us needed to keep a level head and, given the circumstances, it would have to be me. Generations of British ancestors had carried an entire empire on their stiff upper lip. Surely, I could do the same?

—

Dinner was delicious, a seafood bisque followed by grilled, orange-glazed duck and green vegetables and a nougat sponge to die for. We had it delivered to the room to be able to talk and bring Daisy up to speed. Sadly, no one was in the mood to appreciate the fine cuisine. Despite his bravado on the beach, Chris prodded at his food until he dropped the cutlery with a clatter and wiped his mouth.

"Can't eat more. My stomach's knotted up."

"As is mine. Have my sponge, Daise."

"Nah, I'm not hungry either. I'm so sorry for you two. Why don't we do something useful, like start on the photos?"

Chris raked his fingers through his hair and rose. "I guess we'd better."

He headed for the black leather settee where mine and Daisy's tablet computers sat. Not sure what Chris had done to them, but they were running much faster now. His own IT paraphernalia he had packed into a bag on his lawyer's advice, who claimed the cops would search his belongings anyway, but Maître Kerluac would ensure they obtained a warrant first. By having everything in one place, it wouldn't be quite as easy to "accidentally find and search" individual items without authorization.

Petty floated in from the balcony and showered us with rose scent. "Sorry, little one. No time for romance. We've got work to do. Actually, it would be great if you could hang around."

Chris looked up, his long, slim fingers suspended over my

tablet. "What for?"

"Think hexing attack at the lifeguard hut? If there's anything to be seen in the pictures, I want her to check it."

"Ah, okay. Makes sense. Damn, I wish I'd not wasted time on the beach. Or with dinner."

"Chris, hindsight is a wonderful thing. I, for my part, wished I'd come earlier."

When my voice hitched at the last word, the frown returned to his forehead. "Please don't fuss. Can't cope with it. Now let me do one last thing." His fingers swooped and tapped some more.

A sharp retort rose in my throat, but I swallowed it down. Quarrels wouldn't help anyone.

Chris finished his tapping. "Voilà. I've downloaded the app connected to my watch. Like that, you can easily scroll through the pictures on your own computer. No need for cloud links—"

"Hang on, won't the cops be able to trace—"

He compressed his mouth into a flat line. "Trust me, they won't. That's exactly my point. Now, let's see what we have here." He flicked a finger at the screen, where a succession of images was blurring past.

"Blast it," Chris said. "There's a ton of them. It'll take ages to find the right ones. If they're there in the first place."

"I'll give it a bash, okay? Pass me my computer, please."

He did and then slumped in his seat, raking his fingers through his hair when a heavy fist knocked on the door to the suite.

My heart missed a beat.

Chris looked up, his face gray and drawn. "Rats. We're out of time. They're here."

"Monsieur Christopher Lentulus?" A muffled voice shouted outside the door. "Ouvrez la porte. C'est la police. Nous avons un mandat d'arrestation contre vous."

Chris jumped up. "They have their arrest warrant," he whispered. "Quick, hide your tablets. We wouldn't want them confiscated by accident." At the door he shouted, "Un moment, je dois m'habiller."

That was a lie. He was dressed and had packed another bag with clothes, but the flics didn't need to know.

"Deux minutes," snapped the voice outside the door.

"Cinq," Chris snapped back. "Je suis nu."

Even I understood that one. No, he wasn't naked.

Daisy rushed into the room, carrying his gear. "You need anything else?" she stage-whispered.

"No. Here, take my watch." He passed it over the same moment Petty hovered into sight, dipping her washing bowl on Daisy's and my tablet computers. Among a rustling of leaves, she maneuvered the bowl under my hand, still holding the watch. For once, my brain seemed to be fully charged, and I understood her meaning in a flash.

"Great thinking, Petty."

I placed the tablets on top of the soil and hid the watch under her leaves. Loaded with contraband, my familiar then floated outside, where she winked out in an instant.

Daisy appeared on the doorstep, windmilling her arms. "You ready? There's a lot of whispering going on outside. They don't sound happy."

"Nor am I." Chris sighed, and we looked at each other.

"I'll get you out," I said. "By fair means or foul."

A diluted copy of his smile blipped for a second and was gone again. "If you mean magic by that—go right ahead. With pig-headed, uh...pigs after me, I need all the help I can get." He held out his arms, and I was only too willing to hold him tight and breathe in his scent, never to let go.

"Monsieur?"

Gently, Chris slipped from the embrace, picked up his bags, and opened the door on a trio of uniformed coppers, one of whom read him his rights and then escorted him outside. At least they didn't use handcuffs. The remaining cops marched into the suite and rummaged through the drawers and cupboards without once acknowledging Daisy and me.

Whatever they were searching for, they didn't seem to find, for they marched back out, not even bothering to close the door.

"Quick, to the balcony," Daisy whispered.

I followed her outside, but the police must have used the trade entrance, and I didn't get my last glimpse of Chris. He was gone.

Now it was up to Daisy and me to solve the murder.

And Petty, of course.

Chapter Seven

Judging by the dull ache spreading in my jaw, I must have been grinding my teeth without noticing. With Daisy's hand on my shoulder and Petty hovering by my side, I stumbled back into the suite. Once I'd slumped on the settee, my familiar tickled my cheek with her blossoms.

That did it.

All the craziness of the past few days and the pent-up anxiety of the last couple of hours crashed over me, and I lost the plot. Clinging to my cousin, my familiar on my lap, I plunged into a pitch-black abyss, and it took half of eternity to claw my way back to the surface.

"Tissue?" A white square fluttered into my vision, and I used it to blow my nose. Dumbo would have been proud of me.

"Go to bed, Myr," Daisy said with determination. "It's been a day from hell."

I drew the heel of my hand over my eyes and sniffed. "Not before I've sifted through these wretched pics. Otherwise, I won't sleep."

"Then we'll search them together. Where's the link Chris gave you? And what about passwords?"

There was no need for either. The page was still open on my tablet. Eyes glazing, we sifted our way through confused shots of skies and sand, but the camera thingamajig had also taken some surprisingly clear photos. A Super-U delivery van rolling past the beach. A seagull, its wings outlined in the pink glow of the rising sun. Followed by more blotchy and confused images until—

"There's the hut!" Daisy shrieked. "Oh, and there's a shadow, look."

"Shhh. Goodness, if we keep this up, the management will chuck us out. Criminals and mad Englishwomen aren't good for business."

"Sorry. If that's not a shadow by the lifeguard hut, I'm not Daisy Amanda Coldron."

It wasn't amusement I sensed, but something not too far removed either. "You're right. At least this one's clear enough, even if it's upside down. Someone's standing exactly where the red cross is, and where I had this disgusting vision. Petty, care to sneak a peek?"

My magical flower tilted her bowl at the tablet. One leaf reached out and tapped the image. The leaf snapped back. In the next instant, Petty shook her foliage and blooms, as if a mini-twister were raging in her chartreuse washbowl. Then she stilled. Her leaves drooped, and the angry pepper scent boiled up. With a jerk, she slipped aside, launched herself into midair, and there she hovered.

Daisy and I looked at each other.

"Uh, right?" Daisy said.

"She's upset about something. Now we need to work out what's bothering her. Petty? When you're done, do you mind coming back here?"

The primula crash-landed on the table, kicking off my phone. My luck was in since it bounced onto the rug, not the slate-gray tiles. Gazillions of questions weighed on my head. How to choose the right one, in a way Petty could answer?

"That's a person next to the hut throwing a shadow in the photo, correct?"

Yes, she rapped.

"The killer?"

Petty shook her leaves and wriggled her pot in indecision.

"Rats. She doesn't know. But that shadow, or rather the person who threw it, worries you."

One rap. Yes.

Daisy toyed with the tip of her braid. "Because of the dead guy on the beach?"

Two raps. No.

No? Why no? The dead guy on the beach was the real problem here. "Something to do with Chris?"

Two raps. A fresh wave of the angry pepper scent.

Why another no?

"See, we're doing this mostly for him. I need to spring him from the slammer. And we must find the true killer."

Yes, she rapped and sent up a wave of rose scent. Jolly good, we seemed to agree here.

Daisy dropped her braid and leaned over. "If we identify that person, will it help Chris?"

Yes, Petty rapped. But it didn't sound as energetic and determined as before.

"You're not totally sure. You only think it will help."

The single knock rang out in the suite.

At least we knew where we stood. "Who the heck is this person? Okay, okay, stop drooping your leaves. I know you can't answer that one. And you, Daise, don't give me the stink eye."

She tossed her head, looking mutinous. "Wasn't doing that."

I clawed at my scalp, but it didn't help an iota with the dull throb that had been building up. "Sorry, a severe case of shot nerves."

"Hey, don't take it out on me, okay?"

The throb segued into a twinge. The tears were back. I wouldn't be able to keep this up for much longer. "Are there any more photos?"

"Good question. Let's see." My cousin flicked through assorted images featuring white spots and liver-colored blurs. In the third from last, the shed showed clearly. The shadow, however, was gone, swallowed into the bright flare of the rising sun, like Chris said. Two more overexposed images, and that was the end of it, the watch's camera had stopped its recordings.

A thought slunk around the edges of my consciousness but vanished into the undergrowth before I could catch it.

Sunlight. Something to do with the sun. With tiredness dulling every fold of my gray matter, I held on to the one thing that made sense. "Here's our proof. There's another person, most likely the same someone who left the beastly vibes I sensed today."

Petty shot out a spark. It landed on the last clear image, the one filled with light and a bit of the hut. The angry pepper scent boiled up once more.

Daisy wriggled her nose. "I wish she could talk."

"She does. As does Tiddles. We're just too stupid to understand them."

"Tiddles is pretty good at getting her meaning across."

"That's because she has very simple messages. I want food. Give me food. Give me more food. Petty here is trying to tell us something important, and we're not digging it."

A spark, soft and warm like a kitten's paw, landed on my hand.

"I love you too."

"See, that wasn't too difficult."

"Emotions are fine. Information is where we seem to get our wires crossed. Right, one step at a time. I'll zoom this photo to the lawyer straightaway."

"Do you think they'll release Chris?"

"Not immediately, no."

"I feared as much. How long will we have to wait?"

"Wait? I have no intention of waiting. I'll see what other dirt I can dig up. The more ammunition for the lawyer, the merrier."

"Such as..."

"I want to have a word with the woman who reported Chris to the police."

Daisy gulped. "Myr, you can't threaten her. I mean—"

"Daise, seriously, do I look like a Mafia don? I'll tell her who I am, what's happened to my man, that I'm freaking and want to know what she's seen. She can't do more than slam her door in my face. What I'd really like to do is search the office of Monsieur Poussin."

"Ooh, cool. Shall we have a go tonight?"

I couldn't help it. I had to laugh. "No can do. I'm running on fumes. Let's case the joint tomorrow and see if there's any chance of getting in there. If he's installed a security system, the place is off limits."

The lawyer supplied with plenty of photos and a terse message, I sank into the large double bed, Petty parked on the side where Chris should have been. Hyped-up and over-tired as I was, I expected to toss and turn for hours, but somehow I dropped into blessed oblivion the moment my nose hit the pillows.

—

Sunlight washed through the east-facing windows, and I blinked into the rosy brightness filling my room. Sunlight, shadows. Yesterday's inspiration teased my memory. The blaze was important. However, the thought crumbled away, leaving me with nothing but a nagging disquiet. I crossed the arms behind my head and stared at the smoke alarm on the ceiling, but the inspiration was already miles away and running fast.

At least my head had cleared. While my synapses didn't exactly pop—it'd take a gallon of coffee for that—they were back online.

I threw back the covers and picked up my phone. No messages from the lawyer or Chris. Instead, I had plenty of messages from my friends back home, assuring me everything was hunky-dory and no new letters had arrived.

Yay. For a good twenty-four hours I'd shoved the puzzling envelopes to the bottom of my personal dung heap. What else did we have here? Ah, Alma, asking where my Himalayan primula was.

Urgh. Stupid of me. While the Simpkinses wouldn't spot one missing book on a shelf bursting with reading matter, they *would* notice the missing plant.

I shot off a quick message, claiming the primula required special care and that there were hothouses for that. Phrasing the mail in a way that would make them think I'd arranged for primrose sitting without telling them took ages, but I hated lying. The Simpkins sisters deserved better. Also, I didn't want them to think I wouldn't trust them with plant care.

My mail sorted, I took a shower. Like the soapy water glugging around the drain, my worries circled around Chris. Was he up already? How would he be feeling—well, the answer to that one was shitty—and would they treat him right?

I yanked the faucet shut and fumbled for the towel. The steam wafting through the small bathroom reeked supremely outlandish. I reached into the stall and checked the gel I'd dragged along.

Cucumber and sage.

Note to self—Don't buy that again. Nor the gummy bear one.

Having braced myself for snide remarks and knowing looks, I headed downstairs for the breakfast bunker. There was no better word for it. Underground, with a low-slung ceiling and no windows, the place drew echoes from the smallest noise, probably the reason the guests were talking in hushed tones. No one took notice of my arrival, so I filled my plate with renewed confidence and sat.

Joy. The crispy, flaky croissants, rich with butter, melted in my mouth, and the confiture de cassis was a fruity masterpiece.

What will Chris be having?

Hungry no longer, I pushed aside the bread basket. Since it was empty already, the gesture was pointless.

"Morning, Myr." My cousin shuffled up to the table, carrying another bread basket and wearing rumpled pale gray linen trousers and a matching indigo tunic with a complex pendant on a black string. The bruises under her eyes told me she must have slept a lot less soundly than I did. The poor girl always took ages to adjust to new beds.

Feeling guilty for dragging her along, I pointed at the pendant. "Is that new?"

"Yup. Bought it yesterday. Amethyst and silver. It's supposed to bring good luck."

"We need plenty of that. Hope you're ready for a bit of sleuthing."

She perked up. "Oh, yes. Where do we go first?"

"Madame Guillou. I'd love to hear her version of the statement."

"Who told you—"

"Sarah happened to drop the name."

She held up both hands, her fingers splayed. "Gimme ten, and I'm with you."

Back upstairs, I had a missed call from an unknown French number. The lawyer? My heart skipped a beat. Stupid, stupid, why didn't I take my phone down with me? Other people were married to the damn gimmicks, but not I.

I listened to the voicemail.

"Hi, Myr." Chris sounded tired. "I'm allowed one call to you. It's mostly okay, but as expected, they've confiscated my equipment. Now they're trying to get a warrant to search my stuff. Otherwise, don't worry. I've stayed in worse places during my studies. Let's hope that lawyer gets in touch soon. Take care and do the right thing."

Needing to hear this voice, I kept replaying the message until Daisy knocked on my door just in time to prevent another meltdown. After having asked Petty to guard the assorted hardware, we headed out into an overcast day where a gunmetal cover of clouds matched my mood.

Daisy looked up and pursed her lips. "We better take the car."

If I didn't move, I'd soon climb the nearest wall. "Nah, come on. Yesterday, we sat in the van all day long."

Halfway through town, a fine drizzle set in that found its way under my hood, plastering the hair to my forehead.

"Yuck," said Daisy. "I'm getting soaked here."

"Think of it as moisturizer."

She tossed me a glance past the corner of her hood, a deep burgundy that clashed with her auburn hair something rotten. "You're quite cheerful today, given...eh, well."

"I'm not, actually, but at least we're doing something, and that always helps."

For a moment, she fell silent. Then, "You're like Mum. Always proactive. Never take no for an answer. You'll go down fighting. Not like me at all."

A little-girl-lost tone had crept into her voice, which triggered my inner alarm bell. "Not true. We might be different, but you, too, are a fighter. You just do things differently. My default mode is icebreaker, and that doesn't always get me where I want to be."

Daisy sniggered.

To mollify my cousin, I'd chosen the way along Carnac Plage's principal shopping street, which sported plenty of cute boutiques I too would otherwise have loved to explore. Daisy, however, was in her element, fluttering in and out of the various establishments like a wound-up butterfly until she stopped at a shop window.

"Ooh, look here, Myr. Found it yesterday but didn't have enough time to visit. They're selling fab soaps and perfumes."

"You own plenty of perfume, Daise."

And I don't need more shower gel.

"But this is French perfume bought in France."

About to ask what would be so different if she bought her scent here, my gaze fell on a wrought-iron chair placed to one side of the Savonnerie's display window. A garland of roses snaked its way from the backrest to its legs, under which nestled plenty of plastic boxes tied with pink ribbons. That wasn't what had drawn my attention.

I stepped closer. Was this...

It was. Most of the plastic boxes contained gaudy potpourri made of hibiscus or lavender and roses, but a couple of boxes at the back sported aged petals similar to the ones that had spilled from my envelope.

Potpourri à l'ancienne, the inscriptions on the boxes read in a curly script.

"Hah, I recognize those." Daisy mirrored my thoughts as she dabbed at the shop window, leaving a tiny smear. "We could, like, have a word with the owner?"

"We will eventually, but I'm not holding my breath. I imagine they sell that stuff by the dozen. It's worth a try," I said rather hastily, when the all-too-familiar pout appeared on Daisy's face. "Not right now, though. I'd like to catch Madame Guillou before she sits down for her lunch. Sorry, but Chris is more important than this bloody letter business."

"Oh. Okay." A guilty expression had sneaked onto my cousin's lovely face.

Attention span of a gnat, I thought, but not without affection.

Watery sunlight glittered on the sea when we arrived back at Plage Beaumer, where yesterday I'd walked with Chris. This time, we didn't go down to the beach, where the tide was in, rocking the boats, but followed the road curving around the sunlit bay. Only five of the houses seemed to have a view of the bay, and of those, only one bore the name Guillou.

"Shall I?"

"Rather you than me," Daisy said. "I'll wait for you back in the street."

"You speak proper French."

"But my nerves aren't as good. I don't feel like having doors slammed in my face."

"Me neither, believe me. Oh well, it's all for a greater good."

With that, I lifted the brass knocker—a long, slim hand holding a ball—and banged it on the door.

Chapter Eight

Among much snapping of bolts and rattling of chains, the door slitted open, and half a mistrustful face showed in the gap. "Oui?"

"Madame Guillou?"

"Oui?"

What had I been thinking? What was I even doing here? This could only go wrong. The carefully construed French phrases dropped from my mind, leaving behind a total blank.

"Eh, do you speak English?"

The owner of the face snorted. "I was an English teacher." She pronounced it Een-gliesh, but otherwise the pronunciation wasn't half bad. Better than my French, anyway.

Like an opera singer poised to launch a high note, I filled my chest with air. "As was I, but that's not why I'm here."

"What do you want, Madame?" She wasn't budging an inch, but if she was living on her own, it made sense to watch out for odd visitors popping up on her doorstep.

"There was an...uh, incident on the beach a couple of days

ago where you called the police."

"Oui?"

Super, we were back to single syllables. I sucked in more breath to launch my plea. "The man you spotted on the beach that day is my partner. Eh, I mean the man who's still alive. He's now been arrested for something he hasn't done, and I'm doing my damnedest...best to get him out of prison."

Either I had shocked Madame Guillou into speechlessness, or she was pondering her response for absolute ages. Then, she said, "Un moment."

The door clicked in my face. Clicked, not slammed. Presumably, this was a good sign. And she'd asked me to wait, not to remove my sorry backside in a hurry.

Was she calling the cops?

No, the bolts were already snapping and the chain clattering, and when the door swung open once more, it did so fully. A slip of a woman in beige slacks, a blouse dotted with purple peonies, and a matching cardigan stepped out. She blinked at the sunlight and pointed at two wooden benches next to the entrance.

"Sit there," she said with enough severity in her clipped voice to make me feel sorry for her former pupils. But then the French education system was a lot stricter than ours.

I sat. Bad idea. Small as she was, Madame now looked down on me. But that might have been the desired effect. It seemed to help. Her rigid posture unbent. "Eh, where are my manners? You like a drink? Coffee? Tea? The English always drink tea? Or a Calvados, perhaps?"

That made me smile. "I'll have whatever you have, but please, you don't need to do anything. After all, I'm intruding on your privacy."

"I saw what I saw."

"I'm sure you did. The question is whether you saw everything there was to see."

Her haughty brow arched. "Now you catch my interest,

Madame. Wait a moment. We must have a drink." With a determined tread, she returned to her house. This time, the door remained open.

Daisy appeared at the top of the road that led to Madame Guillou's mansion, and I gave her a thumbs-up. She wriggled two fingers, mimicking a person walking, and pointed at the beach. I nodded, and my cousin vanished from sight.

From within the house rang the jingling and clattering of cups on saucers, from which I deduced the Calvados had stayed where it was. It bode ill for my sleuthing skills that, when she reappeared with a tray, it bore not only a cafetière and cups but also a chubby green bottle with a golden label and two wide-bellied glasses.

Oh well, when abroad, do as the locals do.

Madame Guillou placed the tray on a metal table next to the bench and distributed its contents. "I don't eat a big lunch. Like you English, I have a hearty breakfast, and by this time I need more coffee. Where are you staying, Madame...?"

"Oh, sorry. I'm Myrtle Coldron. We're at the Auberge du Dolmen."

"Bof," she said, most likely indicating her displeasure with our choice of accommodation. She poured coffee into the cups, then an amber liquid that slid from its bottle into the glasses like so much oil. "By we, you mean your partner and yourself?"

"He arrived first. Originally, we were planning a vacation. When the police went after him, I traveled across with my cousin and...in my van. Bit of a rushed affair, I'm afraid. I had to leave my bed and breakfast in the hands of my housekeepers, but they're super-competent." I was babbling, a sure sign of stress. Rats, I'd nearly mentioned Petty. A magical zombie primula wouldn't go down well. Calvados or no Calvados, Madame would pinpoint me for a lunatic and throw me out.

"Didn't you say you were a teacher?" Madame Guillou shot me a sharp glance.

"You used the past tense, and I did the same. These days, I

run a business, which I inherited from my aunt when she died, but I was a grammar school teacher before. English Lit and German conversation."

"Ah," Madame sat on the bench next to mine. "Literature isn't for me. Too much confusion over the interpretations, no?"

She handed me a cup from which wafted a delicious aroma. "Mh, smells wonderful."

"Turkish and Jamaican, I have it blended for me. Here, we have a fantastic shop."

Having sipped the fragrant brew—I couldn't enjoy it, not with my nerves coiled like this—I returned the cup to its saucer. Here was my cue. "There seem to be plenty of wonderful shops in Carnac, but right now I'm not in the mood for them."

"Ah, your young man. No doubt you wish to hear what I observed on that day?"

I gave her my best take on puppy eyes, though it wouldn't fool this woman. She would help me if it suited her, not because of anything I might want. "Please, that would be most kind. He hasn't done anything. He was simply in the wrong place at the wrong time."

"Mh." Madame Guillou emptied her cup and lifted the cafetière. She refilled both our cups and sat back on the bench, blinking at the pale sun.

"I will tell you what I told this inspector. A very pushy man. One who thinks that wearing a uniform and being under sixty makes you a more important person than an old has-been like me."

Nudged by instinct, I blurted, "You can't be more than seventy. These days, that's nothing."

She beamed. "Evidemment. But the inspector, he asks the leading questions. He has his theory, and he wants me to confirm it. Tells me I must speak the truth and nothing but the truth. But I wonder..."

My chest ached with foolish hope. "What?"

With a determined *snick*, she replaced her cup on the

saucer. "Let me tell you, and you can judge for yourself, eh?"

—

The shadow of a seagull flitted across the table, followed by a sharp gust. The sun might be out, but the wind had a bite to it that made me snuggle into my jacket and wrap my hands around the warm cup. Madame Guillou in her thin cardigan was made of sterner stuff; she never once shivered. Straight as a rod, she sat on her bench, watching the bay beyond her sea wall, where the boats were swinging on their ropes, dipping and rising on the ripples that feathered across the water's surface.

She pointed at the beach beyond. "From my living room, I have a good view of this part of the bay. That's where I was the other day. Figaro was hungry, so I rose to feed him. My black cat," she explained.

"Oh, you too have a cat. Mine's called Tiddles, a tortoiseshell, but she's a geriatric."

Any trace of mistrust vanished from Madame's slim face. She shoved a strand of her gray bob behind her ears. "Ah, people who own cats are good people. Though it's the cats who own us, no?"

"Oh, yes."

"He is always hungry. So, I look out of my window and notice these two people at the top of the steps. I saw them because they were moving oddly."

"Oddly?"

"Yes, at first, I think they are dancing, no? But they don't do that. I realize they're arguing. The smaller is pushing at the taller. Shaking the fists. The window was open, and the wind carried their voices. I couldn't hear what they were saying, but it sounded angry."

"You said one was taller than the other? Did you notice anything else? Like their clothes, for example? Male or female?"

"It was still rather gloomy. The man who died later was the taller person and blond. The other was wearing dark clothes and moving around a lot. I'm not sure about their sex."

My stomach twisted. Chris always wore black. He wasn't a small man, though. Quite the opposite. If the dead detective was taller, he must have been a giant.

"I wondered what they were doing. But they were too far away to see without glasses. I don't use them much, you know? Seventy-three and, in the distance, I can still see almost as clearly as in my youth." She nodded to herself.

The judge was out on that one. Madame struck me as being someone who prided herself on her wits and general fitness. Bully for her, but not if it meant she got her testimony wrong and my Chris into trouble.

"When I return from the kitchen, one person is lying on the beach, near the steps. The limbs are all twisted, and there's blood on the head. I noticed because of the light hair. And because I wore my glasses. The other is bending over the body and looking around as if trying to check for witnesses. This is the moment when I call the police."

She snapped her mouth shut. Testimony given. Duty done.

"How long did you take to go to the kitchen and back?"

"Oh, not long. Maybe a few minutes? The glasses were not where I left them. They never are." She gave an embarrassed laugh behind her hands and downed her Calvados.

To be polite, I took a sip. The sharp, fruity alcohol stung all the way down, pushing tears into my eyes. Somehow, though, the pain got the mental juices flowing.

How long had she really been absent? And at what time in the morning did Madame hit the brandy? I shook myself. The woman had been most helpful, and here I was, thinking nasty thoughts. But I had to think them. Chris's freedom depended on me doing so.

"Thank you. There is one question I must ask—and please don't hate me. Are you sure the person you spotted arguing and the one who was bending over the body are the same?"

She dipped her chin in a crisp nod. "You ask the right

questions. That inspector didn't. I told him I was torn, but he insisted there were only ever two men. Well, I told him I wouldn't swear the first was a man. Bah. He kept right on insisting. In the end, I agreed they might have been the same. To be honest—" She bit her lip and sipped from her Calvados.

"Yes?"

"No matter what the inspector thinks, I wonder where the hat disappeared to."

I straightened. "Which hat?"

"Well, the smaller of the two people arguing on the beach wore one of these weird hats. Like the fishers used to wear? Dark as well. They look idiotique, if you ask me."

Enlightenment dawned. "Oh, you mean a bucket hat."

"Exactly. The dead man, the one with the light blond hair, didn't wear a cap or something. Anyway, the person bending over him was dark-haired. To me, he looked like a man."

"Sounds like my Chris."

"Yes, and he wasn't wearing a hat. Now, the inspector said, if it ever was there, it fell off during the argument."

"Let me guess. They didn't find it."

"Non."

I looked out on the bay, at sun glints on water, boats and buoys dancing on the swell, and the seagulls wheeling over the peaceful scene.

Not that it was truly peaceful. The shallow waters bore witness to a crime. A crime that Chris had never committed. Madame had just proven it to me. However, as bad luck had it, the evidence was lost, most likely washed away.

I swung around to face the Frenchwoman. "My partner would rather be dead than wear a bucket hat. He hates them with a passion."

Madame scratched her nose. "He sounds like someone I might like. And I swear to you. I can tell the difference between hair and hats even without my glasses. Since there was no hat and your—Chris, you say his name is?—has dark hair, monsieur le flic decided I was a bit gaga and mistaking things. But I know what I saw."

She crossed her arms in front of her flat chest.

Sudden heat surged into my chest. Oh yes, there'd been another person around, but the bloody police, not satisfied with ignoring the evidence, insisted on twisting the witness's statement to suit their theories.

—

Having promised Madame Guillou to keep her in the loop, I made my way down to the beach, where I found Daisy standing at the waterline, gazing at the sea.

She swung around. "How did it go?"

"Madame is a nice person and has been most helpful. Unfortunately, it looks as if the local coppers have made up their minds about who killed that guy, and they're more than willing to warp the evidence if they feel like it." I filled her in on the hat story.

"Can they do that? I mean, ignore clues and frame people like Chris as suspects?"

"They shouldn't, but it happens. A while ago, I read about a guy who was convicted on a rape charge only because running more DNA tests would have been too expensive. He spent over ten years in jail for a crime he didn't commit. Can you imagine? Nor is he by far the only one. Oh, Gods, Daise, this is a nightmare." A black wave of despair rolled in. I sat in the sand and buried my head in my hands.

Daisy crouched next to me. "Come on, Myr, chin up. This isn't like you. I know it looks bad at the moment, but we'll get it sorted. Promised."

She meant well, but my optimism was missing in action.

I sighed. "I guess the bloody lawyer needs to hear about the hat."

"Did he get in touch?"

I checked my phone. No messages. "Do me a favor, Daise.

Can you ring his office and make an appointment? Today, if possible. I can't keep on waiting until hell freezes over. He needs to yank his bum into gear, otherwise Chris will never get out." Yes, it hadn't even been a day. No, I wasn't having it.

"Sure."

"Great stuff. Here, take this." I handed over the sheet of paper where I'd written down Maître Kerluac's details.

"Merci." With thumbs that seemed to have more joints than mine, she entered a number and pressed the phone to her ear.

Her French was halting but clear. It was also clear that whoever she was talking to tried to fend her off. Whenever my cousin wanted something, she got it. She also was a lot nicer about it than I would've been.

Call over, Daisy sat next to me. "Spoke to his secretary and mentioned Madame Guillou. We're on tomorrow at eleven. She claims her boss is on the case and is making them sweat over the warrant. It's a murder charge, so the facts aren't straightforward."

"Tell me something new."

"Looks like that lawyer guy's doing his job."

Was he really? Did I want the impossible?

"That's terrific, but what worries me here—we've already got quite a few bits and bobs that don't add up, and Kerluac should use them to go on the offensive."

Daisy shrugged. "Maybe it's the French way of doing business."

"Huh, I thought they were all about l'attaque. Oh, well. Guess we'll find out tomorrow." While I might have succeeded in sounding brave, I had no clue how to get through the next twenty-four something hours.

Daisy picked up a handful of sand and let it trickle through her fingers. "That's not the only thing I discovered."

"Huh?"

"At the hut I had, like, funny vibes?"

I swung around. Daisy's cheeks were flushed, and there was

a glitter in her eyes. The stiffening breeze had torn strands of hair from her braid that blew around her face.

"Funny as in witchy?"

"Probably. You know I'm not the greatest whiz with the paranormal. But I tell you—something's wrong with that shed thingy. It feels, how shall I say? Slimy? Nasty?"

"Spiteful?"

She nodded. "Yes, exactly."

I smiled at her. She smiled back.

"Hey, great stuff, Daise. We'll make a proper witch out of you yet."

She rolled up her eyes, bared her teeth and growled.

"Now you look like Grumpy Cat."

"Oi, don't diss me. By the way, that wasn't all. Though it might not be a big deal."

"Heavens, what's up now?"

"Look at this." She dug into her pockets and retrieved a wad of crumpled tissues. With delicate fingers, she picked at the package, revealing withered pale-yellow rose petals. Not dried and pale like the potpourri à l'ancienne, but fresher, from a rose that must have been in bloom not long ago.

"Found them stuck under the door to the shed."

"Which is where?"

"Right next to where the red cross is."

Right next to where my skylles hooked on alien thoughts. My gut cramped. Could these petals be the true source of Daisy's and my visions?

We swapped worried gazes. "It might mean nothing," Daisy said. "I used a tissue to pick them up and received no nasty sensations whatsoever."

Her skylles being feeble, she might well have noticed nothing. But what about me? "Okay. Let me try." Not wanting to do this, I extended my digit, tapped the tip of a wilted petal—and pain lanced my finger. A split-second later, smugness slimed into my being, followed by glee much stronger than the experience yesterday when I'd been with Chris.

Unfair. Hate you. Serves you right. This wasn't a whisper, but a snarl.

I jumped back. "Eek."

Daisy's eyes grew enormous and round. "You sensed something?"

My stomach dropped like a proverbial lead balloon. "Yes, it's the same effect as before, only more intense. Which means..." With my mind spinning around and around in a dizzying vortex, it took a while before a picture emerged. "Rats. You know what? These aren't normal petals but manifested magic. Which means—"

"The person Chris noticed was a witch. Oh my gosh, Myr, imagine—a murderous witch."

I pinched the bridge of my nose. I'd feared the same, but we were going way too fast. "Hold it, will you? Sarah would tell me off for seeing nonexistent connections."

We both stared at the tissue and its wilted contents.

Daisy shook her head. "Myr, this can't be a coincidence. We both had a paranormal experience, yours stronger than mine, and now we've found these petals. The Guillou woman sees another person on the beach. Chris sees a person at the shed. The photos show a shadow. A man is dead, and Chris didn't kill him. What does this tell us?"

"That we don't have the full picture. Okay, I suggest we do two things. We return to the petal shop, and we must run these sorry specimens past Petty to make sure they really are manifested magic." I pointed at the wad still cradled in Daisy's hand. "Tonight, we try Poussin's office and see if we can find ourselves more evidence before we visit the lawyer. Actually, it might be a good idea to split up. Why don't you go to the shop, and I'll check with Petty?"

"Sounds like a plan." As one, we sprang up from our crouches. I shoved the packet with the rose petals into my pack, and after a brief hug, we left for our respective missions.

Chapter Nine

B ack in Chris's suite, I unfolded the wadded tissue, and, heart beating into my throat, placed the package in front of Petty's plastic bowl. Too much depended on my floral friend. Chris's safety. My sanity. Everything, really. Like most other human beings, I didn't enjoy being proven wrong. This was an exception.

Please, please tell me these petals belong to a normal rose.

The primula quivered. With a jerk, the chartreuse bowl shot backward. Her leaves fluttered, and a chaotic flurry of sparks shot up. Then she lifted her bowl and banged it onto the coffee table, not once or twice as she used to do, but a whole series of raps that rumbled like indoor thunder.

Ever since my arrival in France, my mood had been on the sour side, but after Petty's verdict, it curdled into a rancid mess.

"Just to confirm I'm not completely on the wrong track—you think these petals are indeed manifested magic?"

One rap. Yes.

An uppercut to my diaphragm couldn't have been more

effective. My legs noodled, and I sank onto the leather sofa, its surface slick and unyielding. "Oh, crapola." And that wasn't all. "Magic left behind by someone who isn't a nice person, yes?"

Another rap.

"I knew it. Oh, isn't that just effing brilliant?" The urge to smash something became overwhelming, but I limited myself to banging my fist on the surface of the sofa. Other than making my tablet hop, nothing much happened. Nor did venting make me feel any better.

I rose to pour myself a glass of orange juice, which I then downed in one go and sat back down on the sofa. A second later, I jumped up once more to pace the room, Petty hovering by my side like a faithful Labrador.

"The only good news, well, good-ish, is that we're right where the magical action is. Or was rather. Someone used their skylles standing next to the shed. And I bet you my favorite shower gel, this potpourri forms part of the action."

An imaginary Sarah raised an admonishing finger. True, I was jumping to conclusions again since there was no proven link yet between the magical petals Daisy found and the scary letters—other than the two having been mailed from Carnac. To top it all, Daisy's and Jenna's missives had given me no weird hunches whatsoever.

An imaginary icy finger prodded my neck. For the murder, the picture was clearer. Four days ago, Monsieur Poussin died, killed by someone he'd been arguing with, someone with magical skylles wearing a bucket hat, someone who had then slunk away to lurk beside the shed, watching Chris incriminate himself.

Unless there had been yet another person running around that morning? Possible, but unlikely due to a distinct lack of clues, so I wadded the thought and binned it.

"To be honest, the letters are iffy, but I could swear there's a connection between the person at the shed, magic, and the PI's death."

Petty's blossoms bounced up and down as if she were nodding.

My imaginary Sarah frowned. Let her.

"Begs the question—what that person was doing?"

Thoughts blew through my head like a flurry of icy snowflakes, bringing back Chris's words. His jitteriness, the urge to rise early. Him turning west when he wanted to go east. "Uh, could it be the unknown French witch used her skylles to summon Chris to the beach?"

The icy finger prodded some more. If that were the case, the French witch might have killed Poussin to frame Chris. But why? What had he done?

Now I wasn't jumping but skydiving to conclusions. My imaginary Sarah facepalmed. Petty sent up a powerful wave of her soothing lemony scent.

"Yes, I know. Imagination's running amok." I grabbed the nearest thought rushing by. "Sweets, do you think the French witch is stronger than me? I mean, *if* she somehow summoned him, she's operating way above my levels."

Petty popped a spark. She then pointed her blossoms at the window, whirling them like mini satellite dishes, which told me she'd tapped into her source of witchy wisdom. After a short while, she shook herself.

No, she rapped.

Oh? Boosted by a sudden lightness, I could have floated to the ceiling. "Wow, I seem to have hidden depths. Really well-hidden depths. Daisy's also doing much better."

Yes, Petty rapped.

I stepped across to the window. Could the endless rows of standing stones, the alignments, have caused Daisy's and my recent achievements in the magical department? Or was Petty being overoptimistic? Even if she was right, I wasn't ready to face our antagonist.

They might want to confront you...

I stomped on my inner voice. It did have a point, though,

and was the reason I called Jenna next and brought her up to speed about Chris's plight and our witchy antagonist. Empathic as she was, she couldn't share calming vibes over the phone, but having access to a sympathetic ear and a person acknowledging my woes flowed a soothing balm over my shredded nerves.

"Sweet Earth, Myr, be careful," Jenna said, right at the end.

"I'm trying my best. Give me a buzz if something happens."

"No worries, I will." She ended the call.

Phew. It had been a challenge and a half, but somehow I managed not to upset her by mentioning our plan to burgle Poussin's office. For which I still needed an open sesame, so I picked up my grimoire.

Then I hesitated.

So far, the recipe book had revealed the secrets in the invisible section after someone died a violent death shortly beforehand. Like a paranormal leech, my magical skylles thrived on murder. On the other hand, what was I supposed to do? I didn't kill Poussin. Nor did Chris. If I wanted him back, I'd have to prove his innocence.

Which was the reason why, a short while later, I was prowling Chris's suite, the recipe book pressed against my chest. Based on experience, this went a long way if one wanted to coax the tome into revealing its secrets. The last time, Daisy played a crucial role, though the very first time I managed the stunt on my own. Emotions always triggered my magical skylles, and the potent mix of fury and frustration churning through my guts should be more than enough for my purpose. Also, I had a clear idea about what information I needed, another important part of the equation.

With bated breath, I placed the recipe book on the table and said, "Recipe book, please show me a hex which can open doors."

I wasn't expecting much. When, therefore, the panels of the grimoire flew open, releasing the musty vanilla scent of old tomes, I jumped backward, my heart fluttering like a caged bird.

Pages flipped over so fast, they blurred. I took another panicky step—and slammed my hip into the sharp corner of the sideboard.

"Ow."

The recipe book calmed down. Pages were folding over more slowly until all movement ceased.

A hand on my smarting hip, I limped across. "Huh?"

The grimoire didn't reveal instructions hitherto invisible. Instead, it had read itself right to the end, where great-great-etcetera Aunt Petunia had recorded the remains of our lore when Napoleon's troops were savaging the European continent. While that might have been a while ago, it was a lot more recent than Lily Coldron's original insights recorded in the sixteen hundreds and afterward hexed into invisibility.

I stepped onto the balcony, studying the text on the open page. "How to change the physical appearance of an object."

No, that wasn't what I'd expected.

I sank into a striped deckchair. My first real hexing stunt had been all about object manipulation, but how was that trick supposed to help?

Unless...

"Ah. I'm to morph the lock in Poussin's office into something else if we want to get in." *Urgh.* That was a tall order.

Trees soughed in the wind. A peloton of cyclists whooshed past, legs pumping.

What did you expect? An easy-peasy spell to go?

To be honest, that was exactly what I had expected. Rats, I should have known better.

Branches rubbing against branches hushed in my ear, morphing into the voice of a dead woman. Rotten she might have been, but when it came to hexing, Jenna's grandmother Dot had been a true ace. "That gun you shifted into a cactus? Myrtle, you were stressed, you were afraid, you were furious all at once, plus you wanted to save my grandson, my great-grandson, and your young man. So, you did."

Dot's grandson and great-grandson weren't in danger, but my "young man" certainly was. Otherwise, I was most definitely stressing for England, while running both scared and furious. When it came to that part, we were doing just peachily.

Would it be enough? When I morphed that handgun, the danger had been imminent. This time, it was more of a cold, dark fear lurking in the recesses of my soul. While I'd come a long way since April, most of my hexing had been more or less spontaneous, paranormal hiccups where I didn't exercise much control.

This stunt demanded tons of focus and determination. Did I have what it took?

—

The door to the suite banged open, and Daisy rushed onto the balcony, waving something square, something wrapped in silk paper.

"Myr, hi. Look here, I bought us some."

My cousin had fast become the sister I never had and, like with any sibling, once in a while we drew sparks off each other. With my patience worn thinner than a Crêpe Suzette, it didn't take much to poke the inner bear.

"Some what?"

Seemingly oblivious of the snarl in my voice, Daisy lobbed the package as if it were a paper grenade. The object certainly was light enough to be just that. As I peeled a plastic cube from its wrappings, an artificial reek of roses and lavender assaulted my nose. "Is that the potpourri? Stuff reeks like perfume de bordello." Now I thought about it, the pungent reek seemed familiar.

Leaning against the railing and raising her face to the sun, Daisy said, "Not sure I would know how a bordello's supposed to smell. Let's see what Petty has to say for herself." She looked down and gave me puppy eyes. "Oh, what about—"

"To cut things short, yes, Petty nailed your find from the

beach. Unfortunately, the answer is we're indeed dealing with manifested magic, meaning we're dealing with a witch. And most likely a killer."

Daisy slumped. "Rats, and there was me hoping... Oh dear, not good, is it?"

"Douze points. Scary, eh? Even scarier, if this potpourri now also carries visions, we really are up shit creek without even a boat."

"Because it would imply Ella, uh...the owner of the Savonnerie, might be the witch. And she's aware of the Misfits."

No, not a ditz at all, my cousin. "That for starters, yes."

Silence fell.

"I quite liked Ella, you know?"

I stared at the sepia-bland mix filling the plastic container. This load featured brown seedpods and cinnamon sticks mixed in with the rose petals. "You sure this is the same product?"

"Because of the twigs? Gimme a mo." She rummaged in a cute turquoise shopper I hadn't seen before. Daisy must have done more than purchasing petals.

"I brought my letter." She fished out an envelope from her new bag and emptied it into the ashtray the same moment Petty floated over, wriggling her leaves.

I flipped the lid off the plastic cube, wrinkling my nose at the odor. "Here we go. If it weren't for those twiggy bits, what you bought looks exactly like the petals in our letters."

"Maybe the witch took out the seeds and things because they distracted from the petals?"

"Hmm. Isn't it a bit odd a supernatural someone would use potpourri when all they had to do was pick up their manifested magic?"

"Did you ever do that?" Daisy twirled the bushy end of her braid around a finger.

"Snap, what a shame I never produced enough manifested magic to feed mass mailings." The faintest of notions that I might be missing something important blipped at the back of my mind. And vanished.

"Maybe our witch has the same problem? Anyway, I asked Ella if she could remember anyone buying the stuff."

"Let me guess. The answer is no."

Daisy pouted. "You're in one heck of a foul mood, aren't you?"

She was right. "Sorry. This crap's getting to me."

"Don't let it. Anyway, you're right. The answer's sort of no. But there's a catch." She raised her index finger, and the braid bounced onto her shoulder. "This type of potpourri doesn't exactly jump off the shelves. Ella likes it because it's elegant and understated, which makes it hard to shift. However, she did sell two packs in the last month, and they both went to the same person."

My heart missed a beat. We'd done it. Found another connection. "Our witch? Unless the woman's lying, of course."

Daisy wagged her head. "True, but I didn't think she was. Shame she couldn't remember a thing about the buyer. Doesn't even know if we're talking man or woman."

"If we assume Ella's being honest, what does this tell us?"

"So far, not much, I guess."

Sadly, my cousin had a point there. I glared at the potpourri doing a successful job of passing for shredded paper interspersed with compost. So much depended on my familiar now. "Okay, Petty. All yours."

My miracle flower hovered her bowl over the mix filling the plastic cubes. Other than a gentle flutter of the pink blossoms, she remained calm. No sparks, no drooping leaves.

To be sure, I asked, "Am I right to assume this load's not giving off nasty vibes?"

She rapped twice for no.

"Looks like Ella isn't a suspect," Daisy said, her tone bone-dry.

"Nope." Without knocking once, my earlier misgiving rushed back into my mind. "Hah, that's it. Why did the potpourri I received in the post carry nasty emotions when yours didn't? Or Jenna's. Or this one. There has to be a reason. That's assuming it's the same stuff."

Daisy tugged at her earlobe. "Begs the question of why mail some of it in the first place."

"A good question. Sadly, I don't have the answer." The words rushed out in a half-snarl. "I still don't know whether I was supposed to stay in Avebury, away from Chris, or vice versa. I might've fallen into a trap."

"And then, you might not."

"No. Yes." My head, overloaded with input, channeled a balloon at bursting point.

"You'll work it out. Ah, did you find a way of getting into Poussin's office?"

Here was something I had an answer to. The balloon in my head deflated somewhat. "Did I ever. The recipe book took me right to the end, to Aunt Petunia's delightful collection of spells."

A frown had appeared on the porcelain smoothness of Daisy's forehead. "It read itself without me being around?"

If I'd looked any closer, I could have seen her thoughts parading along the smooth expanse of skin. "Daisy, the recipe book never stopped at the invisible pages, so this doesn't mean you're not needed anymore. And don't forget. You sensed the magic by the shed, *and* you found the petals."

A smile lit up her face, and the depressed conga line vanished. "True."

Give me strength, someone. "Our grimoire opened on the page that explains how to change one object into another. If I turn the lock into sand, for example, or the door panel into fog, we should be able to get in."

Daisy tilted her head. "Eh, don't get me wrong, but..."

"Am I sure I can do this? No, absolutely not. Even if I did, the effect won't last long enough. I mean, when I messed with the gun, it turned into a cactus only for a brief time. At least I think so."

"How about turning a stone into a key?"

"Yes, but would it be the right key? The instruction says I need to envisage the form I want an object to take. The more precise, the more complicated it becomes."

A frown now crowding her forehead, Daisy toyed with the end of her braid. "Mh. Tricky."

"Very. My only chance is to tinker with the lock, and the moment I change it, you shove the door open."

"Then it will remain open, and the cops will know we were there."

"They'll know *someone* was there. As long as they can't work out who and no one spots us breaking and entering, we should be okay. Emphasis is on should. Heavens, this is quite mad, you know? We don't even know if there's anything worth finding."

Daisy, her cheeks flushed, her eyes glittering with excitement, said, "Oh, Myr, don't be a wuss. You're doing it for Chris."

She had me there, and she knew it.

Chapter Ten

Situated behind a car wash and a Renault dealership, on the ground floor of a nondescript white building, the private dick's office was hidden from prying eyes. At this time of the evening, the car wash and dealership were closed, and the street lay empty. While the lamps pooled their frosty glare on the street, the courtyard in front of the office lay in shadow. Thankfully, the indigo skies were still light enough to make out shapes and structures—such as the narrow gap between the door panel and jamb.

No need to test my skylles. The door was already open.

"Cool. How did you do this?" While the black rayon stocking Daisy was wearing over her head muffled my cousin's voice, it still was much too loud.

I tapped a finger against my lips, squished flat by the twin of Daisy's stocking. "Pssst. Wasn't me." The fabric of my makeshift balaclava did its damnedest to snag on my teeth.

We held our breaths and listened. Apart from the background hum of a small town, nothing but static filled my ears, so I gave the door panel a gentle nudge with one gloved hand. It swung into a corridor, of which I could just about make out the part at the front.

"'Ello? Daisy stage-whispered. "Qui est là?"

Many rapid heartbeats passed without anything happening. Emboldened, she repeated her question, not bothering to whisper this time. The corridor remained as unresponsive as before.

"I don't like this," she said.

"Nor do I. It's better if only one of us goes in, with the other as backup."

We both stared at the now wide-open door.

"Myr, I know I was the one who said let's go for this, but I think we'd better vamoose."

She was right. Of course, she was. Every moral fiber of my respectable middle-class body was buzzing like mad, wanting me to stop this nonsense at once.

Yet we'd come so far. And the open door was too much of a temptation. "Let's try phase one. If someone's waiting inside, they might think we're gone and come out."

"All right."

Phase one was that part of a rather patchy plan where we checked for hidden security cameras in the vicinity. During our earlier visit, we'd spotted one pointing at the car wash, but since it didn't swivel on its pole, it posed little risk.

Not wanting to shock the residents, we yanked the stockings off our faces and retreated into the yard.

No cameras. No flashing blue lights announcing law enforcement or a security detail. No action inside the office either.

Just one open door.

"Looks like there's no security system," I stated, heart beating into my mouth. "Otherwise, whoever opened the door would've triggered it already. Makes sense. I think. Since we're here, I might as well try." Unfortunately, my legs refused to move. This was sheer madness. To gain some time, I pulled the stocking back over my face, and Daisy followed suit.

"What if the witch is in there?"

"I can only hope my skylles are up to the challenge."

Daisy's squished mien was hard to read, but her snort told me what she thought of my plans.

"Daisy, I'm aware this could be a trap, okay? I'll be super careful."

"I wonder whether there's a dead body lying inside."

Urgh. I willed the shadows in the corridor to clear, but they refused to do the honors. "Like who? The owner of this outfit is already dead."

"You have ten minutes. If you're not out by then, I come in."

"It'll take longer than ten minutes to search the office, but I'll pop up in between. You holler if things go pear-shaped."

I took a few hesitant steps into the corridor, covered in light tiles. There seemed to be four doors, two on my left, two on my right.

I inched my hand at the handle of the left door and thrust it open the same moment I jumped aside. No killer burst out, so I pointed the beam of my flashlight at a bucket, a mop, and an impressive array of chemicals lined up on a shelf, filling the small space with an acrid whiff. The same maneuver repeated farther down the corridor yielded nothing more menacing than a grubby toilet. Someone seemed to collect cleaning fluids and not use them.

The left side of the corridor secured, I tried the first door on my right, which turned out to be a spartan waiting room. A quick check of my watch told me no more than eight minutes had passed, but I traipsed back out anyway. "Daise?"

"Here." She peeled from the shadows.

"So far, I've drawn a blank. There's one door left, and I hope that one leads to the office. Give me another ten minutes, okay?"

She fiddled with her watch. "Off you go."

I returned inside, ignored the panicky voice in my head

urging me to run, and depressed the handle of the remaining door. My trusty flashlight confirmed I'd hit the jackpot—one medium-sized office, closed for too long and giving off a stale, sweaty stink, but apparently not containing any burglars or corpses. Nor did I spot anything remotely looking like a security camera.

The office was furnished with a set of metal drawers on one side of the room and a plain white desk facing the store front. The letters arcing across the glass pane created a funny shadow play on the desk's surface, but the padded chair behind was as empty as the wooden one in front.

Since they were closer, I rattled the drawer cabinets first. Unfortunately, every single one of them was locked. No keys, of course, and my skylles were notable by their absence.

I swung the beam of my flashlight across to the desk, highlighting a computer and keyboard, a stack of gaudy filing baskets, most of them empty, a block of square white notepaper, three pens in a beer stein—and a paper desk calendar.

Now, that might be useful.

Ignoring the heat spreading in my stockinged cheeks and the tightness in my chest, I slipped across and flipped pages. A notice saying "CL surveillance" would at least have confirmed what we already knew—that Monsieur Poussin had been shadowing Chris. This calendar said no such thing. In fact, it was pristine. Not a single entry. Wherever Poussin kept his appointments, it wasn't here, probably the reason the police didn't take the thing with them.

I had wasted my time.

Close to me, something clacked.

I froze.

The silence in the room clamored in my ears. My innards knotted.

Quite a while passed until my muscles protested at my rigid stance, so I shifted position—and the zipper of my jacket hit the keyboard, which resulted in another faint *clack*.

False alarm.

I swallowed twice and clawed at the blasted stocking. The fiendish thing turned every breath into a struggle. But I didn't dare to take it off.

How odd the cops didn't confiscate the computer. Would they contract hackers who could crack passwords in situ?

They might. Surely they had his phone as well.

If only Chris were here. Since Chris was the reason for my being here, this thought was of no use.

In the absence of a computer expert, I would need Poussin's password to get in. Fat chance of my ever finding that.

Out of the corner of my eye, I caught a glitter. A foil-wrapped sweet pinned to a corkboard otherwise covered in notes, menus, and photos, none of them of any interest, apart from one crumpled slip of paper.

"PoissonBoissonMarmalade#1962" it read. The word "Marmalade" made my synapses pop. The word phrase reminded me of Chris's outlandish password, even though it was shorter.

No, this was too easy. It couldn't be possible.

What if it is? Could it be that Monsieur Poussin and Chris took their inspiration from the same security manual, only Poussin didn't follow the instructions which warned not to take notes?

It was worth a try.

I shot outside and told a bored Daisy I'd take even longer. Having stopped her from joining me, I raced back inside and fired up the computer.

It booted fast and asked me for a password. With trembling fingers, I entered the code I'd found on the note—and the screen changed to a desert landscape.

I was in.

—

Saved on the desktop, the file stood out like Camembert in a sweets shop. It was labeled Clients Recénts which meant recent customers. About to click on the icon, I hesitated. Chris would know a way to manipulate the file properties, but a lame IT duck like me stood no chance. If the cops bothered to look, they could work out when the file was last accessed.

I'd forgotten the sweaty mask and the tightness in my chest, but they both returned with a vengeance.

I really shouldn't be here.

A good thought, but it didn't stop me from clicking the icon. Apart from four names and addresses, the file was empty.

Whether it was inspiration or yet another fit of madness holding me in its grip, I'll never know, but I also clicked on the file properties and, sure enough, here was the record of my snooping.

That wasn't what made my heart skip a beat.

I took another glance, but the information remained the same. The file had been modified this evening, around nine, most likely by the person who opened the door. What else might they have done?

A warning klaxon sprang up in my mind. *Get out of here.*

I whipped out my phone and took screenshots of Poussin's desktop and the contents of the suspicious document. Once done, I powered the computer down. With that sorted, I rushed outside, closed the office door behind me, and headed for the entrance, while the beam from the flashlight wobbled and wavered over the whitewashed walls and light tiles. Something equally light lay on top of them, something I'd missed on my way in. Various small objects, to be precise.

I pointed my flashlight at a smattering of pale-yellow rose petals.

"Oh, crap," I said, the comment popping out like a shot.

The confused light filtering through the open doorway changed as a shadow moved in.

Heartbeat launching for the stratosphere, I pressed myself against the wall, wishing it would meld. Fizzing started in my ears, and a strange green glow spread from the edge of my vision. The ground under my sneakered feet vibrated, as if it was purring.

My skylles were on the rise, and I welcomed them.

"Myr? You okay? You're like, glowing green?"

Daisy. Not the unknown witch, but my cousin.

Skylles, down.

The fizzing and vibrating eased off, leaving behind the scent of fresh woodlands. I slid down the wall into a crouch and massaged my throbbing temples.

Having unglued my tongue from the roof of my mouth, I said, "That could've gone quite wrong."

"Why? You knew I was there. You've been in and out like a yo-yo, so what's the problem now?"

My muscles were leaden, heavy. Moving my arm meant asking too much. "Behind the door."

"Huh?" Daisy stepped in and switched on her own flashlight. "Oh."

"Exactly. Oh."

"I better pick them up, right?"

"Please be careful."

Daisy, a shadow moving among other shadows, bent over and used a tissue to collect the petals. She hesitated. "Myr, do you think the French witch morphed the lock? And morphed it back afterward?"

"Hence the petals, you mean? It's possible." Nothing this person did would surprise me anymore.

Sitting here was great. It eased my heartbeat, and even the twinge in my temples had calmed down. It wouldn't do, so I hauled myself into an upright position. "I guess we'd better go."

"Uh, why do you think they left the door open?"

"Search me. I suspect it's meant to alert the cops to a break-in. I'm pretty sure they're supposed to find something."

"What—"

"Daisy, later. First, we've got to get out of here."

"Oh. Okay."

Three houses and two streetlights later, I remembered the stockings. We yanked them off and swapped sheepish grins.

"Oopsie. Would've been a scream and a half if someone spotted us," Daisy said.

"Emphasis on the scream."

All the way back to our auberge I was waiting for flashlights, the sound of running feet, and loud voices yelling at us to stop, but that didn't happen. We even made it back in record time.

Over a glass of red wine and some truffles from the minibar—with buckets of adrenaline still coursing through my system, I needed the empty calories—we checked our loot.

One pack of yellow rose petals Petty identified as belonging to our unknown paranormal antagonist and four names and addresses.

"Those names don't ring any bells." I refilled my glass. Hopefully, I'd burned enough adrenaline in Poussin's office that my midnight debauchery wouldn't end up as an extra layer on my hips.

"This one does." Daisy poked the third name on the list.

"Since when do you know a Madame Ella Renard? Oh, she's..."

"Yes. The owner of the Savonnerie."

"But we agreed she can't be our witch. Otherwise, Petty would have reacted to the potpourri like she did to the petals in the letters. Correct, Petty?"

Petty rapped out a quick yes and fired off a spark.

Daisy held a white truffle between the tips of her manicured hands. Glittery nail polish seemed to be back in fashion. This one was purple with a turquoise sparkling effect.

She bit into the praline and licked her lips. "True."

Unkempt thoughts swarmed my mind. I needed some law and order here. Point one—the other three names on Poussin's client list meant nothing to either of us. Point two—none of them was in Carnac, though one person at least resided quite close, in Lorient, while the other two were as far removed as Quimper and Rennes. A proper detective would now be planning a trip to talk to all three of them.

Not me. Or rather, I was planning to start here in Carnac. No matter what she might have told Daisy, Madame Renard was somehow involved; of that I was sure.

My gaze fell on the second, fresh set of yellow rose petals, and an idea swam in from the deep. Before it could sink again, I jumped up and let my fingers slide over the velvety surface of the petals.

And there it was, the echo of the smugness and self-pity I sensed the first time at the lifeguard hut, only less intense and laced with a heavy dose of vengefulness. A quick glimpse of Poussin's office, chased by the inside of some sort of shop, washed over my retina and was gone again.

My fingers stirred the petals, so innocent, so soft. "Makes me wonder..."

"Mh?" Daisy nibbled at a cherry truffle.

I snapped my hands away and fetched a sani-wipe. "Magic of stronger witches manifests as petals. Interestingly enough, they also seem to carry a trace of the personality that manifested them. Together with an impression of places they've seen."

Daisy blinked. "Wow."

"Yes. When I received the first load of petals, I suffered ocean visions and sensed some sort of glee or vengefulness. Well, something along those lines, anyway. Our pale friends here"—I pointed at the tissue—"triggered a similar response. It's less intense, but I just saw the office we burgled and a shop. Also, I captured the same type of spite. Our witch seems to have serious issues here."

"What with? Or whom? Poussin is dead."

Like an old-fashioned pinball machine, levers in my brain pinged thoughts in all directions. "Which means he won't ever access files again. Wouldn't the police notice the document was changed after his death? And not by Chris."

"Don't get your hopes up," Daisy said. "With les flics anything can be possible."

I went out onto the balcony to stare at the night. One after the other, lights winked out as people went to bed, but my body was humming like a swarm of angry gnats. "I wonder about that file they found on Chris, the one accusing him of hacking. Maybe it too was planted."

Daisy rose and stretched. "Stop fretting. It's late." She gave me a peck on the cheek. "Nighty-night, Myr. Try to get some sleep."

"Daisy?"

"Yes?"

"Thanks for being here. With me. It means a lot."

The sudden smile made her face glow. "Anytime." She waved and was gone, the door snicking shut behind her. A few minutes later, I was in bed, but it took hours of tossing and turning before a blissful darkness sneaked in.

Chapter Eleven

Fuzzy shapes flitted through my mind, threatening to yank me into full wakefulness. I wished for them to shift their sorry selves, but dreams never work like that. The shapes solidified, became a real place, a place I knew better than the inside of my head.

It was night, and I was walking along the upper corridor of the Witch's Retreat B&B, using a flashlight. Now, why would I do that? There were some perfectly fine light switches. The beam of my torch was slipping over them. However, instead of turning on the spotlights in the ceiling, I headed for the curtain separating the corridor from the landing, which led to Daisy's and my bedrooms.

My gloved hands splayed over the handle. That was idiotic since I'd locked up before I left.

Mumbling drifted into my ears. It didn't sound like my voice. An instant later, the handle and lock bubbled, melted, and turned into a stocking. Not any stocking, but one of those decorating fireplaces all over the UK at Christmas now hung from a hole in the door panel.

I'd never been much of a Christmas person.

With the lock transmogrified, the door no longer barred my way, and I pushed it open.

Now, why would I hex my door instead of using the key to get in? And why the heck would I be wearing gloves? Maybe this was a practice run for my next burgling stunt. Dreams were supposed to help us deal with problems we encounter during the day, right? However, tossing my own bedroom made little to no sense. Still, I rummaged through my drawers and flung clothes aside in the wardrobe, obviously hunting for something, though I didn't know what.

This dream was pointless. I wanted it gone.

Instead, I was now returning to the corridor. The stocking shimmied, and a split-second later, the handle and lock were back.

I rattled the door a couple of times. It was shut and wouldn't budge.

Filled with satisfaction, because the hex had worked, and frustration, because I didn't find what I'd come here for, I sneaked back down to the entrance hall. Over the coconut runner I crept, headed for the living room and another closed door.

Again, I didn't use my key. Again, the stocking appeared where a lock had been an instant ago. I excelled at this transmogrification hex—

Realization pounced. I didn't excel at this hex. I didn't need it either, not at home. Which meant the person sneaking around in my B&B wasn't me.

Was this even a dream?

With trembling fascination, I observed the intruder as she— yes; we were talking female; I could tell by her slender hands— slipped into the back of the living room, into the den doubling up as office space.

As round as the full moon and just as bright, though a lot

wonkier, the beam of the torch traveled over the walls, highlighting shelves, a photo of Daisy and me as children, and finally the old Victorian rolltop desk. The beam froze until the searcher crept closer, keeping the flashlight focused on the desk.

There *was* a secret compartment, but these days it stood empty since everything of value was in the bank vault. One key I'd taken with me, while the other two remained with Jenna and Rosie.

Good luck to you, whoever you are. You'll find nothing—

An ear-splitting screech shredded the quiet in the room.

It was chased by a menacing growl.

The ray cast by the flashlight wavered and came to rest on something scruffy and small. Something with eyes flaring with a green luminance as the flashlight hit them. Glistening fangs opened, and another deep-throated growl built up in the monster's throat.

Tiddles.

If I needed any proof the intruder wasn't me, here it was. My geriatric pet might grumble when I didn't fill her bowl fast enough, but she never, ever acted up like that.

The French witch—it had to be her—took flight. Flashlight wobbling, she raced into the living room where a pale rectangle of light spread from the window to the floor, telling me dawn was on its way. And an ambulance. Now, how was that supposed to fit the picture? Its blue light strobed across my vision, while its siren yodeled away.

No, that wasn't a siren. Rather, it was...an alarm clock. Rats, I mustn't wake up. I needed to pursue the shady figure.

I was out of luck. The figure misted back into the darkness from where it had come.

—

With bleary eyes, I sat up in my hotel bed, staring at the rosy rashers of dawn grilling on a French morning sky. My befuddled gaze searched the room until it snagged on my phone. The blasted thing had buzzed its way across the nightstand's slick surface, which meant I must have received a call that my imagination morphed into a siren, adding blue flashing lights for good measure.

I fumbled for the phone and tapped on the screen. Bingo. One missed call, three minutes ago. The UK number on the display appeared vaguely familiar.

My heart hopped in my chest. Chris? Since the cops had confiscated his phone, he would have to be released from jail to call. Please, please let him be free, I pleaded with whatever guardian angel was on wonky witch detail.

Sweaty fingers slipping over the surface, I thumbed the number and listened to a ringtone brr-brring away.

Come on, pick it up.

"Myrtle?" asked a brisk voice.

Not Chris. Alma.

I slumped against the headboard of the bed and swallowed a groan Alma didn't deserve. "Good morning, yes, it's me. You rang? What time is it at your end?"

"Just gone six. We haven't been here long."

"It's early enough."

"Sorry for disturbing you, but I reckoned with you being an hour ahead, it would be okay."

"Sure, what's wrong?" The Simpkins sisters could take anything in their stride; for Alma to call at such an uncivilized hour, something serious must be afoot.

"Tiddles was having forty fits, hissing and spitting like water on oil. Cec's given her food, and the poor dear has calmed down. Though she was in a right state. The bit with the stocking is the weirdest of all."

Uh-oh. Realization slammed its Doc Martens into my guts. Not a dream, but a vision. Something cold slimed down my spine. Once again, my skylles were doing things while I wasn't fully in the driver's seat. As if that weren't enough, it appeared our witchy antagonist, having succeeded in luring me to Carnac, was now back in Avebury.

"Stocking?" My voice sounded high and squeaky.

"Yes, through the door to your parlor. There's a hole where the handle and lock used to be. Looks as if someone lasered both out. It's that precise. And there's a Christmas stocking shoved into it. Nice one. Brand-new and good quality as well. Yesterday, we gave the place a good dust-up. No stockings, I tell you."

Like frightened mice, my thoughts skittered all over the place. Our witchy antagonist must have been in one heck of a panic to forget the stocking. Tiddles deserved a tin of her beloved tuna.

Make that five tins.

"Was anything stolen?" I knew there wasn't, but I had to go through the motions. Alma would expect such a question.

"Not that I noticed, no."

"Who's staying at the Witch's Retreat right now?"

"Hm, the Belgian family returned home yesterday. Right now, there's the two Druids in Five and Six and a Mr. Winterbottom, a cyclist, who turned up on spec and is now in number seven. Oh, and Randy Johnston also popped up without a reservation. Very unlike him. Number one was free, so he could have it."

Like Flora Saddler and Iris Shuttlecock, Randy Johnston was one of my regulars. His and Mr. Winterbottom's unscheduled arrivals would have been suspicious had they been females. Unless I got things wrong and saw a man in my scary non-dream?

No, I didn't. The fingers on the handle didn't resemble a man's hand in the slightest. And the vibes were somehow female. The Belgian mother, then? Belgians spoke French, and the two countries were neighbors. Since the family had left, that theory was way out there, though I might have to keep her in spec. Of course, the witch might not even be staying at the

Retreat. But she seemed to know her way around my place.

Blast it all.

"Did Iris call? Or Flora?" To the best of my knowledge, neither of them had ties to France.

"No, why?"

"Just wondering, never mind. To be honest, I'm stumped." That was a gross understatement.

Alma snorted, "No wonder. How's Chris, by the way?"

"He's in real trouble, and I have to look after him."

"Trouble with the coppers?" Her voice was mild, but it didn't hide the keenness for juicy morsels to share with her friends.

I'd have to share something otherwise she'd never let go. Not the fact that Chris had been arrested for murder, though. Even if our churchwarden watched French news, it might take Anna some time to hit the jungle drums. "Unfortunately, they seem to like him as a suspect. We have a lawyer, and there's clear evidence he isn't guilty. It'll take a while to get things sorted, though. I can't leave him, you understand?"

"Sure. If anyone can keep him safe, it's you," Alma said in soothing tones.

For some odd reason, her comment triggered this morning's second lightbulb moment, and the explanation for the recent horrors became all too clear. Oh yes, I knew what the beast wanted. And I would warn Jenna and Rosie to ensure it remained hidden.

I gnawed my lip. That might not be enough.

"Call a locksmith and get the door fixed." While it wouldn't keep out the witch, it would stop the other guests from strolling into the room. Petty and the recipe book were safe with me, but I didn't want auntie's ashes to have an accident.

"Will do. See ya. Everything will turn out fine with your Chris. You'll see."

Triggered by her kindness, my eyes swam with sudden tears. "Thanks, Alma. Take care."

"You too." She called off.

Cripes. My to-do list was growing longer than the queue at the boulangerie. Even warning my friends back home might not be enough. Yet, how was I supposed to be in two places at once?

I looked up. There was no need for that. We still had the petals from the shed and the fresher ones I found at Poussin's office. If the vague theory burbling along in my gray matter was correct, touching the petals, the essence of the French witch's magic, had somehow connected me with her, calling up the vision. There was no reason to believe the trick wouldn't work twice.

"Petty is it true witches can use personal items like nail clippings for nefarious purposes?"

My plant floated from the windowsill to the nightstand. One firm knock. Yes.

"That explains why she mailed potpourri in the first place. She wouldn't dare share the essence of her magic."

Another yes.

"Odd that I sensed her spitefulness when I touched the potpourri, especially when Daisy's and Jen's missive gave me nothing." A ripple blew over the contents of my stomach, and I had my answer. "Urgh, for this much rancor to seep through, I guess I must be topping her personal shitlist."

Rap.

Great, just great. And that wasn't even my biggest problem.

My thoughts swirled into a spin. Since Petty and the Coldron grimoire were here with me, our antagonist was most likely after the magical plaques of the witches, two inconspicuous Neolithic tablets which might or might not be keys to the underworld. Or something else entirely. If not that, there still was our magical mirror, the antidote we needed each year to fend off the curse.

Extreme measures were called for.

With Petty watching on, I opened the recipe book and headed straight to the section at the back. If I found nothing in there, I could still try the invisible text. My luck was in, and I struck gold in the part not studded with wonky skulls, Great-great aunt Petunia's favorite warning sign. "How to protect what is yours" sounded perfect for my purposes, and the hex as such was easy-peasy.

Okay, the fire alarm was a thing. I searched my butt pack for the candles I carried with me ever since a three-day blackout had ruined a holiday and slipped onto the balcony, where I singed the witch's yellow petals to a crisp. As the smoke wafted away, I called up an image of a woman in my mind, lifting the magical plaques and then twitching back her hand, covered in a red rash...

Hang on. You'd better not let her get that far.

"I want her to break out in a rash whenever she *thinks* of stealing what is ours," I said to the wind. A green buzz rushed away from me, carried on a woody scent. Then nothing. No pain, no headaches, which meant no magical penalties.

Since I didn't want to kill, only to stop her, that made sense. Still, I waited. With magic, one never knew. As the minutes crept by, a leaden tiredness settled on my limbs, so I lay down on the sofa and closed my eyes. Long-distance cursing wasn't for the faint-hearted.

—

I must have snoozed off, but the squeals of a rogue jazz saxophone bursting from my phone yanked me out of my stupor. I sat up and thumbed the call open. "Hello, Jen."

"Just so you know it—you went to see Rosie ten minutes ago."

"I did what?"

"You asked her for the key to the bank vault since you'd forgotten yours in France."

For a moment, I stood there, unable to utter a single word, my mouth agape. That didn't happen very often, and it didn't take too long either. "I *am* still in France. Oh, blast. Did she hand over—"

"Tut-tut. You should trust us. She sensed something was wrong and slammed the door in your face...eh, the person impersonating you."

"Oh. My. Gosh. You know what this means?"

"Yes. Our French foe can take on other people's appearances. For them to pull off such a stunt, they must've nicked something belonging to you."

Oh?

"Hmm, she—I'm pretty sure we're dealing with a woman, by the way—was searching the Witch's Retreat at the crack of dawn." I briefed Jenna about my dream. "I sort of popped up mid-act, and to my mind that happened because I touched her manifested magic."

"As she did with something of yours. A few hairs are enough."

Note to self—lock your brush in the safe.

"What tipped Rosie off that it wasn't me?" The phone slipped through my sweaty fingers, forcing me to change hands.

"It looks like she can't change her size, and she seems to be quite a bit shorter than you. Also, she came across as...being blurry. Plus, she must've had some sort of rash. She kept scratching her arms."

A giddy rush fizzed through me. Yay, I had nailed the bitch.

"Rosie reckons the witch chose her because neither Damian's nor her eyesight is great these days. Fun fact—she has new contacts."

"Three cheers for Rosie."

Jenna's laugh tinkled like a bell. "Yeppers. And there's more. Once she'd secured the door, Rosie peeked through the guest toilet window."

"And?"

"Our foe flew off the handle. We're a bunch of bastards, especially you. We're responsible for everything wrong in her life. She'll show us, or words to that effect. Oh, and she'll get our plaques. The whole time, her face and body were shifting and changing. But Rosie couldn't work out who's behind the facade. Only that we're most likely talking female, as you said."

I shuddered at the imagined visuals. "Revenge. I knew it. How good was her English?"

"Rosie thinks that if she hadn't known the woman was French, she wouldn't have noticed. Sounds like her voice is a tick robotic but otherwise not a bad imitation of yours. Anyway, that was the first excitement of the day."

I groaned. "There's another one?"

"To be honest, this one's old news. Bob Ignatius tried to get hold of you yesterday. When he couldn't, he darkened the Colonel's doorstep close to ten p.m."

"I messaged him I'd be unavailable. He was told not to show up unless there's an emergency."

"There is. He's received another letter. Couriered from France and warning him against cooperating with us. He expects you to sort things out like yesterday."

"To heck with him."

It must have been the swearing that kicked loose a thought. "How can she courier letters from France when she's in the UK at dawn the next day? And yes, I'm sure it was dawn. The sun wasn't up yet."

"Courier the letter in the morning and bust a gut to get across?" Jenna suggested.

"Not when she broke into the office of Monsieur Poussin around nine p.m. yesterday evening, leaving more petals behind."

"Monsieur Who?" Jenna asked.

"The guy Chris is supposed to have murdered."

Shocking surprises seemed to work wonders for the system. Quite a few pieces of the photo puzzle snapped together, and I shot from the sofa. "Dawn. Sunrise. The shadow. Of course, that's it. I knew there was something wrong with those pictures."

"Eh, you're not making much sense."

"Some photos Chris took when he found the dead guy are overexposed. He thought it was the early morning sun, but it can't be. The shed is facing west. Even in France, the sun rises in the east, so there's no way it would show behind the shed."

"Huh? I'm not sure I'm with you."

I could have wailed with frustration. To me, it was crystal clear. "In the picture I've passed to the lawyer, you can see a shadow, thrown by a light source in the east. That must have been the sun, rising where it's supposed to be. This business with the overexposure—what if it's caused by magic? Jenna, what sort of magic comes with strong light effects and forces a witch to leave her manifested magic behind? Think petals."

"Light effects? How did they look like?" Jenna's voice had gone flat, as if something squeezed the air from her throat.

Uh-oh.

"Super bright and all sparkly. The image isn't the greatest, but it must've been one heck of a blinding flash."

Static buzzed into my ear. "It's got to be... No, that's impossible. We're not capable of such stunts. Though it would explain how she...Oh, sweet Earth, this isn't good news."

Talk about confusion. "What isn't?"

"Teleportation."

The comment rammed straight into my gut. For quite a while, we both remained silent.

"Jenna?"

"Yes, still here. That was a bit of a double whammy. Wow. Think advanced magic."

What I'd done was laughable by comparison. "You're telling me she can flit between the UK and the continent at her leisure? Can she also travel in time?"

"No." Jenna sounded determined. "Space only. Even the Whites couldn't time-travel. Believe me, teleporting from one place to another is a big thing. Though she must be new at it. Otherwise, she'd never leave petals behind for you to pick up."

"Rookie or not, teleporting sounds scary. Makes me wonder how I ever managed to zap the woman."

"Zap her how?"

"I...uh, I was afraid she might be hunting after the plaques and the mirror, so I jinxed her to break out in a virulent itch whenever she even thinks about nicking something which belongs to us."

"Direct magic attacking another person? Are you nuts?" Jenna screeched into my ear.

I couldn't blame her. Like my aunt, I was straying into danger territory with my eyes wide open. "Calm down. Everything's under control."

"Oh, really?" Her voice had lowered, but not by much. We were talking serious stress here.

Join the club.

"Fine, then it isn't. But I couldn't exactly sit on my hands. Note this—I didn't go on the offensive. Nor did I suffer fallout symptoms."

"No headaches and stuff?"

"Nah. I turned rather sleepy, but that was all."

"Phew. Seems like your purpose is justified. But, Myr, please be more careful. Though Rosie said the witch was scratching like mad, and maybe that was your hex. And maybe it messed with her glamor."

Pride welled up. I'd done it, had fought back without breaking any of the cryptic rules and regs governing magic. "She'll try again. We must protect the plaques and our mirror."

"Forget the latter," Jenna said. "Given the woman's still alive and hexing, she must've gone through the ceremony that allows us to carry on for another year. She'll have a mirror."

"She might want to stop *us* from going through the ceremony next year."

"Oopsie, true. Anyway, we organized a coven meeting yesterday, and we're agreed she's after the plaques plus something from Ignatius's secret library, otherwise why bother with him? The petal letters shook us up. The trouble with Chris forced you to leave for France, and bang, here she is, on Rosie's doorstep. When did you say you'll be back?"

"Hopefully by tomorrow evening. We're meeting the lawyer in about an hour. Once that's done, we'll have another chat with the owner of the shop where that beast bought the potpourri she used for her mailings."

Jenna gasped. "Sweet Earth, yes, she couldn't exactly send us her own petals."

"Which she then leaves lying around."

"Told you she's got to be new to this teleportation business. Oh, and she's most likely flying solo. Otherwise, she'd have an Igor equipped with dustpan and broom."

Warmth flushed my chest. "No coven to give advice and back her up."

"Nope."

A memory of the warmth glowing in my body, my being, I ended the call. The French witch, as strong as she might be, had taken on an entire team of magical misfits. She would regret that.

However, she'd also killed a man and landed Chris in jail. High time I had a word with his lawyer.

Chapter Twelve

Maître Kerluac's office was a study in beige. Ultramodern, it featured muted ceiling lights, shelves inset into the walls, porridge-colored leather and chrome furniture, canvases each displaying a single black brush stroke, and lacquered, pale wooden floors. Even the slim blonde receptionist who offered espresso and asked us to wait "un moment" was kitted out in the muted tones of clouded coffee.

The lawyer himself, once we were escorted into his inner sanctum, chameleoned into his bland surroundings. He was also a lot younger than I'd expected. Mid to late thirties was my guess. His thinning, mousy hair framed a lean face, its most noticeable feature a pair of baby blue eyes, gaping at the world in permanent bafflement.

My mood plummeted. This was the man on whose skills Chris depended for his freedom? He wouldn't frighten a rabbit.

"Mesdames, sorry to keep you waiting," Kerluac said in a booming bass that belonged to a much larger and more substantial man. He then flicked through a file and nodded to himself. "Ah, yes. There we are." He tapped his nose. Cleared his throat. "Eh bien, the situation is delicate and very complex, I'm afraid."

His English was fluent with a charming accent. His dithering rubbed on my frazzled nerve ends.

"I presume you received the photos? They clearly prove there was another person besides Chris at the beach on the morning of the murder. He's innocent. Oh, which reminds me. What about bail? I want him out of prison."

An infinitesimal shake of the head glaciered my mood. "Désolé, Madame. It's not easy. There's sound evidence—"

"Lies and conjecture." I was beyond rude, yes, but I couldn't help it, not when Daisy, Petty, and I seemed to be the only ones bothered about Chris.

Kerluac coughed. "Later, we shall look at this. As much as I regret it, bail won't be possible. Monsieur Lentulus is a foreigner, so..."

You little twerp. "Listen—"

He raised his hand. "Please, I'm on your side. I'm having your photos analyzed—"

Blood shot into my cheeks. "They're genuine. Who do you take me for—"

"Madame, these days, anything can be faked, even the truth. When I approach the police, I want to be able to prove this isn't the case here. I must be honest, Mesdames. It doesn't look good. The magistrate, she has studied the evidence, and she's sure there is a case."

Kerluac droning fanned the flames of my fury.

"Of course, there's a case," I snapped. "A man is dead who deserves justice. However, Chris didn't kill him."

Kerluac leaned back in his chair and narrowed his eyes. Where he radiated fuzzy vagueness before, his poise had segued to watchful, and there was a glint in those baby-blues that hinted at another persona behind the harmless facade.

"Myrtle talked to Madame Guillou," Daisy said in her

breathy little girl voice.

Kerluac tilted his head.

"See, she's adamant she spotted someone wearing a bucket hat arguing with the victim," Daisy continued in the same hyped-up tone. "She fetched her glasses, but once she'd returned, Bucket Hat was gone, and instead a dark-headed person was bending over Poussin. In other words, Chris. She's not convinced the two strangers she saw were the same. And the hat has vanished."

The glint in the lawyer's baby blue eyes turned glacial. "According to the police, the witness insisted she only saw two people."

I fidgeted in my chair. "Chris hates bucket hats."

"This might well be, but couture isn't evidence, Madame."

Idiot.

"Madame Guillou insists the police tricked her into saying she saw a scene she's not sure she saw."

Kerluac's brows arched. "Is she willing to make an official statement to this effect? Madame Coldron, I'll have to verify this, you understand? The photo evidence is so-so." He waggled his hand. "However, if there's also a problem with the witness, this becomes, as our American friends say, a whole different ball game." His indifference gone, Kerluac's whole body radiated keenness.

I leaned in. "It means the police are willing to bend the truth to nail their suspect. Tell me, Monsieur. Do you think Chris did it?"

"Mea culpa. I was guilty of that crime. It doesn't matter. I will always defend my client to the best of my abilities. I am renowned for it. Feared even. I simply didn't appreciate that my client chose to lie to me, or so I thought. I'll admit, when I sifted through your pictures, I began to wonder."

About time as well.

"Bon, I will verify with the witness. If it is as you say, this case is not as straightforward as it appears."

"The file, Myr, tell him about the file," Daisy said.

"I was coming to that."

"File?"

"Poussin's. We visited his place yesterday, and, uh, found the door open."

As one does late in the evening.

This time, one side of Kerluac's mouth kicked up in a lopsided grin. "Naturellement, you had to verify nothing was amiss? Ensuring you wouldn't leave fingerprints or be seen, non?"

"Indeed. It appears someone accessed the victim's computer and manipulated a document. However, what exactly they did I can't tell you. Chris could probably work it out. Not from a prison cell, though."

"May I ask how you know someone accessed the computer? Or rather, how you managed?"

"Password on the pinboard."

Kerluac huffed. "Stupide."

"Very. Convenient too. Monsieur, did the police ever notice someone broke into the office yesterday?" Daisy asked.

"They seem to have received a warning concerning a break-in, but when they searched the place today, they found nothing. Apart from an open door. Since they'd already...how does one say? Toss? the office before, they weren't in too much of a hurry. Nor do they seem to have been meticulous enough. I'd say, if someone wanted to create more trouble, that person didn't succeed, no?"

"Possibly because they framed Chris too well. The cops have a warm body in lockup. No need to look for anything which might trouble the waters."

Kerluac produced a very Gallic shrug. "C'est ça. If you are right, I wonder why someone might be so keen to see Monsieur Lentulus in prison."

I looked him straight in the eyes. "I couldn't tell you. Though the dirt on Chris the cops found in Poussin's office and the manipulated file I discovered yesterday imply someone can walk in and out of the victim's joint as they see fit. That's another clue I suggest les flics investigate."

Kerluac sniggered. "No, I won't bother with the inspector, believe me. Instead, I'll go straight to the magistrate. I trust this evidence is correct, since I can only do this once. And while I will send someone to grill the witness, getting the computer checked is more difficult, vous comprenez?"

"I understand. But it was modified and voilà, here's the proof." I held out my phone.

Kerluac scrutinized the screenshots and sighed. "Not evidence we can use in court, sadly, since the evidence is obtained by illegal means. Bon, mail me a copy anyway."

"Surely, you can suggest to the magistrate someone should have another go at the PI's computer." Daisy said.

"I will do my best to suggest this without drawing attention to your...eh, illicit presence in the office." He sounded amused. "Anyway, you have convinced me, Mesdames. Now, I have a lovely surprise for you."

"What do you mean?"

"Monsieur Lentulus is asking after you, so I arranged for a chat."

My heart banged out a hopeful double beat. "I'll be able to see and talk to him in person? That's wonderful."

"Non, non, desolé. For the time being, we have to be content with a video conference." He rose, and we followed suit. "You've given me some work to do. Madeleine will look after you."

We shook hands, and I followed the beige assistant into another office.

—

"Thank you for sticking with me." Chris's face was gray, his dark locks unkempt, and a heavy stubble covered his cheeks. Whatever camera he was using couldn't be great; the image was low on coloring as if a drab film covered the scene.

I kissed my fingers and tapped the tired face on the screen. "You look like a thug." My voice rang with over-bright cheer, but that was better than the sobs lodged halfway up my larynx.

He too kissed his fingers and touched the screen. "I feel like one."

"You're not and you know it."

"At least one person who believes in me, hooray. Even my hotshot lawyer doesn't." Chris clawed at his scalp, mussing his hair further.

"Oh, that might've changed recently."

"Hah."

"For obvious reasons I can't tell you what has happened and, more importantly, what will happen. Kerluac will put you in the picture the next time he visits."

"This evening."

"Good. Then you'll know what's going on. For the moment, rest assured we've been busy. Very busy." My inner imp tempted to stick a tongue at the unknown copper monitoring our meeting, I limited myself to grinning at Chris. "Be cautiously optimistic."

He grinned back. "Emphasis on cautiously. Heavens, Myrtle, you're giving me hope. I needed that."

Realization whacked me over the head. Chris didn't know yet our antagonist possessed magical powers. The whole day we'd been busy sleuthing, he'd spent in the slammer.

I needed to be careful with my choice of words here. "Chris, the person behind all this, is...different. Petty was the one to find out. She says hello, by the way."

Did the screen of Kerluac's computer freeze? Chris wasn't moving, was staring at me in slack-jawed shock.

He straightened in his seat. Okay, so the screen hadn't been frozen after all.

"Say hello to her from me, too. She's certain?"

"Is she ever. We've plenty of evidence in different shapes and sizes."

Chris was a highly intelligent man. He would understand I was telling him some of my proof was of a paranormal nature.

He hummed the theme music from the Harry Potter films and tapped his chin as if deep in thought.

There, I'd known he'd work it out in a jiffy.

Chris stopped humming. "This affair doesn't concern me alone."

"More like having to do with me. And your uncle."

"Looks like someone must've thought Christmas, Easter, and the summer holidays all fell together when I showed up on that beach."

"Nah, I don't believe in coincidences. Your urge to jog there was certainly special."

"You don't think..."

"You know very well what I think."

Thoughts rushed through my head faster than the tide. Here was my chance to get my take on events across to the French coppers, whether or not they liked it. "Let's assume for the moment our perp followed you to France and hired Monsieur Poussin to dig up some dirt on you. Did you hear about the argument on the beach?"

"The one between Monsieur Poussin and me that never happened, no matter how often the inspector insists it did? Oh, yes."

How I wanted to tell him putting words into people's mouths seemed to be the inspector's modus operandi, but doing so would only muddy the waters.

"Nothing is as clear as the police would like to have it. Talk to Kerluac. Chris, I might have to return to the UK soon. There's been some developments."

Chris nodded. "I understand."

How much more could I tell him without tipping off the police eavesdroppers about the supernatural aspects of this case? Though they would be on the lookout for clues to a murder, not references to a rabid paranormal being.

"Not that it matters where I am. The person I have in mind can travel great distances just like that." I snapped my fingers.

Chris hopped in his seat as if jolted. Then he shook his head. Shook it again. "That's impossible."

"Quite." I made sure to hold his gaze and slipped him the tiniest of nods.

There was enough tension in his body to fry a whole server farm. "Jeez."

"That's putting it mildly."

Chris turned to the left as if listening to someone. Then he swung back, looking sad.

"Seems we're out of time."

That couldn't be true. We'd only just started. "It was good to talk to you." I swallowed a ball-sized lump in my throat.

"Likewise. Myrtle, be careful. This is super scary."

"Yup, but we'll get on top of this. The entire gang is on red alert."

On Chris's side, someone spoke in a determined tone.

"Got to go. Love you. Speak to you soon."

"No, see you soon. In person."

We were still touching each other's images on the screen when it turned black.

—

By the time I had pulled myself together, Daisy had driven three quarters of the stretch back to Carnac.

"This is getting to me." I wadded another tissue. Hopefully, it would be the last one. "Which is probably the desired effect."

Daisy shot me a frown. "To take you out of action, you mean?"

"At least keep me on my tippy-toes, so I don't bother with whatever else is going wrong. Roundabout, watch it there!"

"Sure." Daisy zipped through the obstacle without blinking once. "But this must mean the witch considers you a threat."

I released my grip on the passenger seat. "Har, har."

"Oh, you're doing great. This rash hex was inspired." Daisy rushed full blast into a thirty zone.

"Ease off, will you? I don't need a ticket on top of everything else."

"Spoilsport." At least she slowed down. "What do you think she might want the plaques for?"

"Since they're supposed to be magical keys, they must open *something*. Unfortunately, I don't know what."

"Didn't you or Chris mention the Whites used the keys to escape through the henge? Words to that effect, anyway."

"Yes, but that can't be what she wants. Seriously, come on. They left centuries ago. Right, Daisy, you need to turn right."

She stomped on the brakes and shot off into the correct lane, cutting into the traffic. Loud hooting told me someone wasn't happy.

"When in France, drive like the French." Daisy blew a kiss at the rearview mirror. From what I could see over my shoulder, the driver behind us suffered an apoplectic fit.

"That maneuver was very French."

"Bof, as they would say here."

Daisy squealed to a halt at a red traffic light, pressing me against the backrest.

French maneuvers indeed. I gazed out of the passenger window at something flat and neon-blue that looked like a Maserati for the financially challenged, rumbling beside us and revving its engine as if it wanted to play games with us.

Hmm, play games with us? Run rings around us, more likely, just like the French witch was doing.

A French witch who fakes appearances and speaks decent English.

The fine hairs on the nape of my neck bristled. I was on to something here. If I only knew what it was.

The guy in the fake Maserati revved his engine once more.

To fake people's appearances, one needed to know them. Or was it something to do with the woman's language skills?

My brain fog cleared.

It was both. How could I have been so dense?

I swung around to face my cousin. "Funnily enough, our French witch might not even be French."

"Because she spoke almost accent-free English?"

"That's one thing."

The traffic light jumped to green, and the car next to me roared its defiance at the road and shot off.

Broccoli brain.

I pulled at my seat belt. "It just came back to me. We keep forgetting something. This woman knows a lot about us. She has no problem finding her way around the Witch's Retreat. She knows where Rosie lives and seems to be aware of her and Damian's eyesight issues. Somehow, she also got wind of the bank vault where we keep the plaques."

"Excellent research? Divination?"

"Possible, but somehow I have a hunch that this isn't it."

"Okay. Tell me what I'm missing."

"That the witch isn't foreign, but a coven member."

This time, the silence in the car lasted until an industrial complex had given way to the woodlands above Carnac.

"No way," Daisy said after a little while. She gave a brittle laugh. "You're making things up as you go along."

"Nope, I'm afraid not. Would explain a few issues, wouldn't it? Such as the motive, for example."

"Myrtle, hello? Reality check. We suck at magic. All of us apart from Jenna and you. The other two competent ones, Mum and Dot, are dead. Who among the sorry rest do you think can look like someone else and teleport from France to the UK and back again just like that, huh?" She snapped her fingers, like I'd done earlier.

"Gloria Mornings. Either her or Emma Bingham. If you ask me, it's more likely the latter."

"Gloria? She of the gnomes? Together with the woman forever harping on about not being able to hex? You must be joking." A sign read *Carnac Plage, 3 Km.* Daisy indicated and turned.

"I'm not. She's definitely upset. Mrs. Mornings mentioned something. As did the Colonel."

"Because training hasn't happened, and their skylles have gone AWOL. Someone who can teleport doesn't *need* any training."

Put like that, Daisy had a point. Yet there was something with those two that rubbed me up the wrong way, like sandpaper on a cat's fur. "They're both troublemakers. Always stirring the pot. Nay-saying of whatever is proposed. They want much more than is reasonable, especially Emma. This constant hoo-ha might well be an attempt to pool the wool over our eyes."

"Mph," Daisy said. "I'm not buying it. They're pains in the neck. But otherwise?"

Since we'd arrived at the parking behind the Savonnerie, I shelved my suspicions. No matter what Daisy said, I was on to something. It would have to wait. Next stop, Madame Renard.

Chapter Thirteen

Our luck was in, and the Savonnerie was not only open but free of customers.

"Ah, take a deep breath," Daisy said. "Doesn't this smell heavenly?"

If heaven featured an aromatic but chaotic whiff of flowers, herbs, and sweet oils, then yes, this was as close to being divine as it could get. Apart from offering a galaxy of scents, the Savonnerie stocked bottles big and small, colorful soap bars stacked in neat pyramids, tiny oils on carousel shelves, open bowls of potpourri, and too many other goodies to appreciate them all. Anything that smelled of anything was in here, clamoring for my attention. For a moment, the shop seemed familiar, as if I'd visited before. But then, I'd been inside plenty of perfumeries and soap shops.

"Can I 'elp you?" Madame Renard asked. Slim, in her mid-forties, she wore jeggings, a long tunic that was striped cream and blue, and nerdy black glasses which managed to look chic. She didn't look much like a villain, but then who did?

I willed my fluttering heart to take it easy and said, "We hope so. My cousin visited the other day and bought some potpourri."

"Mais, oui, I remember. My favorite à l'ancienne. And I remember you." The look she bounced Daisy sizzled with admiration.

Daisy stepped up onto a soap pyramid. "These are great, Madame Renard. Where do you get them from? I run a small shop in the UK, in Avebury, where I also sell soap, but they're nothing remotely as fragrant as these. This is class."

Madame gurgled a laughter. "Please call me Ella. You have a shop as well? Mais c'est formidable. Oh, I absolutely must give you the name of my supplier. It's in Provence, of course. This is soap from Marseille, very special, very good."

I tapped a lilac bar. "Lavender?"

"Lavender and pine. They specialize in mixes. Try this." She held out a bar in a soft honey color that teased my nostrils with the fragrance of lilies and mint.

I stomped on a dumb urge to ask if the soap was available as a shower gel. "So sorry, but we have a few more questions."

She leaned against her counter. "Sure. If I can help you, I will." Her gaze strayed across to Daisy, who looked back at me, a question in her eyes. I nodded encouragement.

"Uh," she said. "It's like a bit weird. Do you know a Monsieur Poussin?"

Madame Renard...Ella tapped her teeth. "No. No, I can't say I do. Who is he?"

"A private investigator."

"No, what would I need one for?" Her eyes wide, she did that shrug the French have honed to perfection. "I wouldn't know where to find one."

If her puzzlement was an act, she could star in the next blockbuster.

"He died a few days ago," Daisy said. "Suspicious circumstances."

"Ah, yes. Now you mention it, I remember people talking, but I didn't know his profession. Does it matter?" She tilted her head like a sparrow and smiled at Daisy, who smiled back.

My thoughts were in turmoil, tumbling over each other in their race to the top. Either Ella was acting and lying about being a client of Poussin's after all, or she had never met the guy. Since Petty had no beef with the potpourri Daisy bought back yesterday, my money was on the latter option. Which meant Ella's name must have been added to the list by our mysterious antagonist.

The big question was why.

While Daisy and Ella chatted about their respective shops, I tried to untangle my thoughts. What did we know? The witch most likely bought the petals at Ella's Savonnerie. Somehow, some of the witch's spite and vengefulness seeped into the petals I received. That woman must have a temper from hell—

Hang on, temper.

"Did you happen to have a customer who acted up in your shop recently? Someone who felt treated unfairly or something? I suspect it might've been the same person who bought the two boxes of your favorite potpourri."

Ella nibbled at the earpieces of her glasses. "Hm." That must have helped, for her eyes widened and she slapped her forehead with her palm. "Mais bien sûr. How do I forget? I do not tell you this last time, no?"

Daisy shook her head but had the presence of mind not to say anything.

"Yes, now I remember. An Englishwoman, like you. Strange accent, but we also have this." She sniggered. "Go to the Languedoc, and you won't understand a word. Alors, this is not what you want to hear. Yes, she was English. Not tall like you, but smaller, older. My shop was full of people. That was her problem. She wanted to be served first. As if she were special. But I make her wait. I respect my customers' needs and wishes, but I expect to be respected back, not ordered around as if I am a...a...caniche."

Since my inner translator threw out a blank, I had to ask. "What's that?"

"Poodle," said Daisy.

"Ah, how rude."

Ella nodded. "Yes, she was. Threatened me, even. I would regret this, she said. I's funny, non? The face I cannot remember. The words, however, I hear in my ear. Bah, some people are simply épouvantable."

Daisy grinned, which told me she hadn't quite grasped the importance of the shopkeeper's statement. "Customers from hell, yes. Don't we love them?"

If the woman nodded any faster, she'd give herself a backlash. Fearing we were due for another exchange about customer service, I asked, "Did you ever meet the woman before?"

"Non," Ella said with determination. "Never, ever. I am sure of this."

My thoughts were churning and tumbling.

Not a resident, then. The witch must have arrived at the same time as Chris or shortly afterward and spotted the potpourri, the ideal prop for her schemes. When Ella didn't tug her forelock fast enough, the witch took offense and devised a way of causing trouble for the shop owner. Only the police didn't follow up on the false lead. Since our foe was over here, she mailed her letter to Ignatius. If not that, the mailing had been her main purpose and the spiteful addition of Ella's name to the PI's list an afterthought.

While I didn't stand a chance of proving my theory, it made sense. Most importantly, it explained a few loose ends. But when it came to nailing our antagonist, we'd hit the wall. Ella and her Savonnerie were a dead end.

—

This being summer, the ferries between the continent and the UK were busy. Since I was raised to watch my pennies, and old habits die hard, I didn't feel like shelling out a fortune for the express service tonight. Instead, we secured tickets for the boat tomorrow and, much to my cousin's delight, visited the beach. The fine sand warmed my buttocks, and the air was heavy with the perfume of summer—tangy sea air and suntan lotion, mixed in with a greasy odor that drifted from the stall selling pommes frites. A seagull stalked past, a malevolent glint in its yellow eye, though that might have been my imagination.

"This is the life." Daisy was wearing oversized, heart-shaped Lolita sunglasses and one of her minimalist bikini masterpieces, this one powder-blue dotted with tiny ice cones. She was lying flat on her belly, her chin resting on her arms, her glistening skin the source of the coconut scent. Not wanting to get lobstered, I'd thrown on a white linen shirt that smelled of Chris and a broad-brimmed straw hat with fake cornflowers I'd discovered in auntie's wardrobe. The evening sun, still going great guns, found the gaps in the weave, and the crown of the hat was already warmer than Tiddles's belly.

Despite the heat, my core refused to unfreeze. My thoughts were with Chris, locked up in his cell, breathing filtered air untouched by the sun and the breeze that toyed with my hair. If I were him, I'd give everything to be sandy and sweaty.

Once we lose something precious, we realize its true value. I picked up a handful of sand and let it trickle through my fingers.

"Stop moping," Daisy mumbled into her arms.

"Am not."

"Oh, yes, you are. I can sense it from here."

She was right next to me. No wonder she sensed my mood. "I wasn't saying anything."

"No need to. You're radiating stress."

I waggled my fingers. "Witchy vibes, witchy vibes."

"Sounds like you suffer from a severe case of the toddler bug. Did Jenna call again, or what's wrong now?"

"No, Rosie did. Our antagonist seems to have gone into hiding, hopefully licking her wounds. Failing that, she'll be plotting her next move."

Daisy lifted her head. "Relax. She can't get to the plaques."

If only things were that easy. "I wouldn't bet on it."

"Even if she takes on your appearance again, surely the bank won't let her in the strongbox without the secret password." With a contented sigh, Daisy sank her chin back onto her arms.

"I specifically forbade them to do that, no matter who told them to. They assured me their procedures would prevent such mishaps." I hooked air quotes with my fingers. "Since I'm such a mistrustful person, I told them someone might impersonate me and made him promise they would ring first. They'll think I've gone batty."

"Forewarned is forearmed."

How I wished life would work that way. Or were my niggles yet another example of what Aunt Eve always called my glass-half-empty attitude? "Hmm. Maybe my jitters are owed to Aunt Eve being the custodian of the magical keys before me. She did absolutely everything to protect them while I'm swanning around on a beach, relying on my friends and a bunch of bankers."

Daisy sat and blinked over the rims of her glasses. She reached for her shopper and shoved it under her neck as if it were a cushion.

"Myr, give it a rest. We both deserve some fun after this recent madness."

I picked up another handful of sand and sieved it through my fingers. A tiny snail shell, fragile like glass, tickled my palm. Gently, I closed my hand around it. "You might be right, but to me, right now, fun feels wrong. I know I'm a control freak. I know the ball is now in the lawyer's court, that Jenna, the Colonel, and the rest of the gang are competent people. Maybe we should've gone home after all."

Daisy rolled her eyes. "Not that discussion again."

"I know, I know. I'm being contradictory." And cranky.

"Yes."

"Avebury is where the witch is, after all. Or at least where I think she is."

"Exactly. You can't get into her head. We discussed that as well."

"I sort of managed once. Maybe I should try again." I grabbed another handful of sand. No shells in this one.

"How?"

"Aye, there's the rub."

"Shakespeare?"

"Yup."

"Fat load of good a long-dead bard wearing tights is to us." Daisy sat and fished a small bottle from her shopper. "Want one?"

"No, thanks." Daisy's drinks tended to be as garish as they were sugary.

She slurped from her bottle. "Ah, that's better. Would you agree with me this place is boosting our skylles? So, it's good we're here, right?"

"Mh." The theory had a lot going for it. Traveling was supposed to broaden one's horizons. Perhaps there was more than a hint of truth in that claim.

"If we stay here long enough, we might eventually be able to teleport home."

I couldn't help it. I had to laugh. "Complete with car? Sounds like we should beam the entire coven over here instead of trying to train them. Maybe that will make Mrs. Bingham and Mrs. Mornings happy."

Daisy grinned. "Nothing will make those two happy."

Not if one of them was our antagonist.

My bag played the jingle from Harry Potter. A green shadow blipped over my vision and vanished when I blinked. As if a garden had gasped, a sweet and musty scent dripping with moistness wafted by and was gone.

"Who's that? Jenna?" Daisy's voice rushed in from far away.

My skylles were rising, and I knew the news wouldn't be good.

I rummaged for my phone. "No idea. The blasted thing keeps switching ringtones. Even Chris has given up on reprogramming it."

"Could be a reservation for the Retreat."

"I've redirected the business calls. Hello? Myrtle here."

"Where exactly is here?" Sarah, brisk and business-like, which meant she was wearing her hat as a police officer. But there was an edgy undertone in her voice that confirmed my suspicion. No, the news wouldn't be good.

"France."

"I suggest you come home pronto."

"What have I done wrong?"

"It's not about you."

Gone was the taste of summer, carried away on the breeze. As the earthy scent rolled back in, a bitter taste coated my tongue, and I knew.

Oh heavens, I knew.

The next instant, I saw.

A head on a pillow, its owner asleep, the face too fuzzy to identify. A white cloud drifted in, hiding the head.

Not a cloud. A pillow in the grip of two gloved hands now thrusting down on the sleeping head. The pillow jerked and bulged. The sleeper must have woken up in a panic and was struggling against the lethal mass.

Arms flailed. A hand cramped into a claw raked over the wrist of the offender. But the hands held the pillow in place, never once letting go.

As the frantic gyrations of the victim weakened, the scene faded. With a sizzle and pop, I was back by the sea.

"Someone's died. When?"

"Early this morning."

Daisy had put away her sunglasses, regarding me with a puzzled expression. My vision wobbled, and my cousin's lovely face was overlaid with Jenna's sweet features. Next came Mel's fluttery presence and Rosie's gentle smile. Though, no, Jenna and Rosie must be alive since I spoke to both of them.

Why should the victim be female? There was Marty, staid and solid, Damian and his thick lenses, fussing over his wife, and the Colonel.

Please, please, not one of them.

What if it was one of the children?

"Sarah, who is it?"

Her laugh was bitter. "That's a first. So far, you've been the resident corpse spotter. You seem to have been offline the whole afternoon. I asked Ms. Wytchett, but she didn't have another number."

Yes, Jenna was still alive.

Please, please don't let it be Mel. Or Linda. Or anyone else, really.

"Sarah who? And how—"

"Suspicious death. Your neighbor, Gloria Mornings."

Chapter Fourteen

From Cherbourg to Poole, the Britanny Night Flyer took only ninety minutes. Still, with the driving and waiting time thrown in, plus a few comfort breaks, our journey home lasted almost eleven hours. After a short and restless night, morning arrived way too early.

"Will Sarah mind if I hang around?" Free of its braid for once, Daisy's hair cascaded over her shoulders in a glorious auburn wash.

"I certainly won't."

The doorbell gonged. Sure enough, here was my favorite police officer, her sage pantsuit crumpled, her short, dark hair a spiky mess. In the house across the street, where Mrs. Mornings used to live, the lights were on and figures in white coveralls flitted past the windowpanes like ghosts of the plastic age.

The scene-of-crime officers were on the job.

A handful of neighbors hung in there, but the stragglers were dispersing, whipped on by Constable Cameron.

It must have been raining. The air was fresh and a lot cooler than in Carnac. The gnomes in Mrs. Mornings's garden dripped moisture, as if the gaudy figurines were bemoaning the passing of their mistress.

She didn't pass. She was murdered.

"Come on in, Sarah. Fancy some breakfast?"

"I won't have the time, but I'll take a coffee. Listen, if you had some for my team—there's four of them—it would be great."

"I'll ask the Simpkinses."

"On it." Daisy strode off.

Sarah stepped into the corridor. "Thanks a bunch. Where can I wash my hands?"

I flapped my hand at the door next to the utility room. "Guest toilet for you and your team. We'll be at the back."

In the living room, Tiddles sauntered toward me and wove figure eights around my ankles. I lifted the cat and nuzzled her plushy fur. "Hello, my little watch kitty."

"She's a fab judge of character." Daisy entered and placed an overloaded tray on the table next to the sofa.

I shouldn't be hungry. A coven member had been murdered, and I hadn't been able to prevent it. Worse, I'd pegged the poor woman for the killer. But after my broken night, I wasn't only as hyped-up as a class full of teenagers, but also ravenous.

A footfall in the corridor announced Sarah's arrival. "Oh, this smells great."

It did. I waved at the sofa and table, where Daisy was spreading out calorific goodies courtesy of Alma and Cecily. "The ladies spotted your car and hit the hob. Help yourself."

Nose quivering like a ferret's, Sarah ogled the fried sausages, eggs, and bacon Daisy was uncovering. "I mustn't. I'm on duty."

"Okay, alcohol is off limits," I said. "Grease and stodge, though? You need something to keep you going, girl, as Mel would say."

With a groan, Sarah sank onto the settee next to me and stretched her legs. "Okay, okay. Five minutes of peace would be nice. Makes me feel like kicking off my shoes."

"Why don't you?"

"Barefooted coppers aren't exactly awe-inspiring. Apropos the police, how's Chris?" she asked with her eyes closed.

"Frustrated, as you might imagine. The lawyer took ages to crank his behind into gear, simply because he thought Chris was guilty and was not willing to admit it."

"Legal eagles are the pits." Sarah's eyes were still closed, but her nose twitched again when Daisy poured the contents of the cafetière into three mugs.

"Did you talk to the French officer once more?"

"Not recently, no. Last time I did, the guy was adamant he's caught his killer. I'm wondering how much of an open mind he'll keep."

"He won't." I filled her in on Madame Guillou's observations.

Halfway through my story, Sarah snapped her eyes open and drilled her gaze into mine. "Witness manipulation? When there's a distinct possibility he's caught the wrong man? My oh my, that won't end well for my French colleague."

I had a tough time lining my thoughts into some semblance of order. How to put Sarah into the picture while keeping the paranormal aspects of this mystery under wraps? Yet, in killing Mrs. Mornings, the witch had forced my hand. I would have to share at least some of my intel to help Sarah nail a double murderer.

A stomach growled, probably Sarah's. Mine joined in with an unhappy burble.

Sarah wriggled into an upright position. "These aromas are more than I can take."

Daisy grinned and passed the plates.

"Ms. Wytchett said you'd have some intel to share."

Oh, she did, did she?

The villain's true nature was off-limits. As was anything that might expose the coven. Therefore, the right spin called for some supernatural storytelling skills.

"Cream?" Daisy asked.

"I shouldn't, but yes, please." Sarah held out her cup.

In desperate need of a caffeine boost, I sipped from my mug. Then, I placed it on the table. "Jenna's right. A knuckleheaded police officer isn't the worst of my worries. Well, it's different for Chris, but the actual issue is that the murder of Monsieur Poussin and now Mrs. Mornings's death are most likely related."

Sarah was chewing a piece of bacon. "Wrrf?" was the only response she managed.

"My thoughts entirely. See, the saga started with some wacky letters, some of them mailed from Carnac, some hand-delivered here in the village. I suspect—and please, at the moment it isn't more than a hunch—that the chief motive for the killings is revenge. And possibly extortion."

Sarah swallowed her mouthful. "You've lost me. Can you please start at the beginning?"

"This is complex." Desperately needing to clear my fuggy brain, I sipped more coffee. The trick was to tell the truth without giving everything away.

Sarah wiped her mouth. "Here are some questions to get you started. Extortion over what? Revenge for what?"

Don't screw this up. "The bit about the revenge is unclear. I haven't the foggiest why this person might get their knickers in a twist." That wasn't quite the truth, but not too far from it, either. "It's complicated, because the origins of our current troubles started way back in time."

Daisy widened her eyes. From Petty, her terracotta pot parked beside the sofa, drifted the tiniest of rustles, but otherwise she remained motionless, patient familiar that she was.

Sarah emptied her cup and fished a tablet computer from her oversized purse. "Go ahead. I'll listen. If you're sitting on information pertinent to the case, I need to know what it is."

"It's no coincidence the village has seen such an influx of new residents over the last two years. My aunt and Dot Wytchett between them assembled the descendants of a group of early environmentalists, the so-called Earth Wardens, here in Avebury. The Coldrons are one of them."

"Oh? You never mentioned that gem before. Right, you're telling me that's a problem for someone because..."

"No idea. It appears the same someone not only dislikes me...us with a passion, but seems fixated on getting their sweaty paws on some relics which belonged to the Wardens."

I didn't look Sarah straight in the eye. She was a pro and would know in a heartbeat I was hiding something. Instead, I fiddled with my watch and blew a strawberry strand of hair from my face. The cobwebby mess needed a cut, but there was no time for treats.

"What's so special about these people?"

Ouch, that was the one question she shouldn't have asked.

"Do you believe in magic?" The question was moot. Sarah believed in facts and nothing but facts. However, magic *was* a fact of life, even if she would never know.

As if to prove me right, Sarah snorted. "You should know me better than that."

There and then the truth stared me in the face. She and I could never be true besties, not when we were on the two sides of the great divide between the so-called world of reason and the realm of the fantastic. While I might be on the outside looking in, contrary to Sarah, I at least knew the supernatural existed.

"The problem is, someone seems to have a different opinion. They're chasing after two Neolithic stone plaques, which my family has been keeping for centuries."

Sarah pecked at her tablet. "Are they valuable?"

"Not per se. Lots of them hanging around in our region. These two specific ones, however, are supposed to be magical objects."

Daisy was facing away from us, pretending to admire the contents of the bookshelf.

Sarah looked up. "Harry Dresden much?"

Now that was amazing. "You read paranormal novels?"

She shrugged. "If I get the time for them. Takes the mind off things. Anyway, someone has a fixation on your relics, because they make a nice ingredient for their latest potion and sends threatening letters to force you into handing them over, correct?"

"To be honest, there weren't any actual letters." Apart from Ignatius's. However, I'd better keep him out of spec for the moment.

"Huh?"

"We received a bunch of rose petals. From a potpourri. I know, sounds crazy, but petals...uh, were important to the Wardens."

"It's amazing that perps still manage to surprise me. Rose petals." She snorted. "If there weren't any letters, how can you know what this person wants?"

"The w...perp surprised Rosie by asking for the keys to the vault where I keep the plaques. When she hit a wall, she freaked. But she was...in disguise, so I'm afraid Rosie can't give you much of a description."

Urgh. That had been a near miss with the witch.

"I'll have to verify this."

Sorry, Rosie. Hopefully, there would be enough time to warn her.

"Do you suspect anyone in particular?"

I swallowed. If I wanted Sarah to stand a fighting chance of catching her perp, she needed to know. "Yes, Mrs. Mornings's best friend, Emma Bingham. Mrs. Mornings gave me the impression her bosom buddy was up to something."

More pecking. "Thanks, we'll check her out."

So would I, as soon as this grilling was over. I needed to do my bit for justice, but I still had a coven to protect.

"What I don't get is how the two victims fit the bill."

Daisy, facing our way again, had relaxed her posture and was curling a lock around her finger. She dipped her chin in the tiniest of encouraging nods.

"Looks like someone framed Chris for the killing. And no, I'm not referring to this idiotic French cop." Despite sitting on my sofa, I could sense the thin ice crackling under my bum. Things like visions, hexes, or magical petals revealing the mindset of their owner wouldn't help much with my street-cred.

"Then what do you mean?"

"Sarah, what if someone broke into the PI's office and planted the folder incriminating Chris? Someone like our extortionist, for example. You said yourself it was way too convenient."

"Mph. Sometimes, I say the silliest things. How can *you* be sure your spoofy perp planted the blasted thing?"

The imaginary ice splintered, and I envisioned black water rising to my neck. "Because there was another break-in, and to have the place burgled twice by two different people is too much of a coincidence, if you ask me. Plus, there's a connection to the petals."

Panicky thoughts zipped through my head. Not enough time had passed between the witch's presence in Poussin's office and Mrs. Mornings's death for the beast to have swapped continents by conventional means. Hopefully, no one bothered to check those file properties. At least Sarah had bought the petal story.

A laser had nothing on Sarah's gaze. "And you know this because..."

"Because we visited the dead guy's office ourselves, okay?" Daisy said. "We know it's illegal, okay? No way could we let poor Chris rot in jail. Not to forget, we found the door open. Like, it's not that we broke in. Myrtle only had a little go at the computer—"

Sarah slumped on the sofa and covered her eyes with one hand.

"No worries," I reassured her. "The password was pinned to the board."

Sarah groaned and stuck one finger in each ear. "I haven't heard this, I haven't heard this." She withdrew her fingers. "Carry on. This one isn't my case, so I'm not obligated to take offense."

Here comes the tricky part. "The person who sold the potpourri was mentioned in a list of recent contacts. So, we talked to her."

"Oh, you did, did you?" Sarah asked in an acidic tone.

Great stuff, she'd swallowed that one. "Yup. And we struck gold. Looks like someone didn't enjoy the customer service, which is probably why she added the name of the shop owner to the client list, thus putting her into the crosshairs of a police investigation. Sarah, this person is super spiteful and is spinning the truth faster than your average blender."

"Killers are seldom nice people. You are telling me this person killed first the private dick and now Mrs. Mornings? Not to forget a spot of outlandish extortion, breaking and entering, planting evidence, and framing Lentulus."

"Yes."

"Just checking. This is wild."

"It is. With Chris, she succeeded. With Ella, that's the owner of the Savonnerie in Carnac, it didn't pan out because the local police didn't react to the tip with the open door. They might have to change that, since Chris's lawyer is on the case."

Sarah loaded her bun with raspberry jam. "Calories, I need calories. This is unbelievable. Are they total morons in France? No, don't answer me. I know they're not."

"Maybe the inspector is related to your boss?"

"Now, there's a thought. Anyway, there's one thing that bothers me more than anything."

The ice rose to my neck. "Yes?" I said in a squeaky voice.

"You used the female pronoun a few times. How do you know our killer is a woman?"

A gelid wave rose and crashed over my head. Me and my glib tongue.

Despite the blooper, all was not lost. Quite the opposite, since Sarah had also slipped on her own patch of ice.

"The owner of the soap shop told us. Oh, and Rosie's sure she was dealing with a woman. Am I right about the killer being female?"

Sarah stared at the ceiling. "I can't believe I'm that stupid."

Makes two of us.

"You're tired. Don't blame yourself."

Daisy hid a grin under her greasy hand. We weren't out of the woods yet, but with Sarah preoccupied with her slip-up, she might not ask more embarrassing questions.

"Yes, I'm tired, but that was a rookie's mistake. Okay, ladies, this doesn't leave the room. Do you understand?"

"Yes," Daisy and I chorused. From my right sounded a muffled rustle, but Sarah didn't seem to have noticed.

"A neighbor couldn't sleep and noticed a dark-clad person slip out of the house in the early hours, carrying a pillow. They insisted on this being a female and thought in terms of sleepovers gone wrong. Anyway, she must have used the pillow to smother your neighbor."

That the witch didn't use magic for her killing made sense. Based on everything I knew about magical laws, the backlash would have been horrendous. Why didn't she teleport from the scene? Instead, she seemed to have slipped across and searched the Witch's Retreat.

I twitched in my seat as if zapped. Teleporting. Something to do with teleporting was important.

Not now.

"I'm surprised the killer wasn't more careful."

"She probably thought no one would be awake and watching."

In a small village, someone was always watching. Unease burned in my throat. Mrs. Bingham, my surviving suspect, had been living in this place for the best part of two years. Surely, she would've known about insomniac curtain twitchers?

"Anyway, to sum this up. You're convinced the would-be blackmailer, who you think killed your neighbor and framed your partner, is also responsible for the dead PI."

"Yes, don't ask me why she killed him, but I'm sure she did. I can share the evidence I gave to the lawyer."

Sarah rose from the sofa. "Please do. Honestly, it sounds as if Mrs. Mornings was killed as a warning, because the killer wants these plaque thingies. As for the motive for the French murder, I have no clue. Yet. I'll see myself out. Thanks for breakfast, ladies."

My relief lasted all of two seconds when the thought stuck in my brain finally popped.

"Oh, no. Now I've ballsed things up completely."

"Why?" Daisy asked. "I thought you did a great job there. Believe me, Sarah'll catch the witch pronto, and Chris is in the clear."

"That was the plan. However, who guarantees the beast doesn't teleport right out of jail again?"

"Ah."

"My thoughts entirely. Oh, drat, this means we must catch her first. And put a spell on her. Just don't ask me which one. Oh jeez, Daisy." I buried my head in my hands.

"Talk to the Colonel. He knows Emma. He can help you trace the wretched woman."

Three cheers for my cousin. I was already thumbing Elmsworth's number.

Chapter Fifteen

Ouagadougou Cottage greeted me with the sweet scent of honeysuckle hanging from the porch. The door, green and shiny, stood ajar, and classical music drifted out.

I knocked on the panel. "Colonel?"

"Come in," he hollered from somewhere in the back. "I put the kettle on for a cuppa."

Tea sounded great, so I headed into the living room, dark even on a bright summer's day, which this wasn't. The traditional landscape paintings and portraits of horses set against gloomy backdrops didn't help. It was a room meant for winter, for chilly days spent by the stone fireplace, restored with great care and equipped with a modern filter.

The colonel, carrying a massive silver tray, trotted in from the kitchen, rattling glasses, a silver teapot, plates, and something looking suspiciously like a chocolate cake.

That was the XXL-version of a cuppa. "Oh, yummy. You needn't have gone all out."

Grinning, Elmsworth deposited the tray on his gleaming

cherry wood dining table. "One can never eat enough chocolate. Grab a pew."

I sat on a brocade chair I suspected to be authentic Chippendale, if its carved back, curving legs, and ball and claw feet were anything to go by. While the chair might not be my style, it looked quite at home in the cottage.

Elmsworth lit a couple of candles and sat. Then he jumped up again. "I'll tone down the music."

"It's fine."

"Nah. For me, Schubert isn't background music. He deserves attention."

To be polite, I asked, "Where's Buster?"

"Asleep. We had to see the vet yesterday because of a problem with a claw."

"Oh?"

"Never mind, he'll be fine. It's good to have you and your cousin back."

"I'm ever so grateful to Daisy. We shared the driving together with the stress, and it made a world of difference. She's in her shop now. Got a lot to catch up with."

"Family should stick together. Okay. I still find it hard to imagine Emma guilty of such horrid crimes."

Straight to the point, that was the Colonel for you. "I find it hard myself. But she and Mrs. Mornings were as thick as thieves. When I last spoke to my neighbor, she voiced...misgivings. Something about Emma acting before she thinks and wanting to talk to her. Now Mrs. Mornings is dead. This can't be a coincidence."

Elmsworth sighed. "Probably not. To be honest, I scoured the list of female coven members but can't for the life of me think of a motive for any of them. It appears Emma's indeed the most likely candidate."

"That's what I told the police."

"You did what?" Suddenly, he sounded rather wheezy.

"I want my Chris back and her behind bars. There's a catch,

though." I briefed the Colonel on my meeting with Sarah and my worries that the woman might teleport straight out of jail.

When I finished talking, he sighed. "A tricky situation. If you ask me, you played your hand well. Though you're right, if we want her to stay put, we must flush out the woman before the cops do and slap her with some sort of sticky hex."

"We can try, but she's a lot more competent with magic than we are."

Elmsworth pulled a face. "I refuse to throw in the towel. She needs stopping, and for that we need all hands on deck."

To my understanding, he had been in the army, not the navy, but I let it stand.

The colonel shoved one hand into the pocket of his beige chinos. "Back to Emma. She and I were the first to move to Avebury, so naturally we talked a lot. Gloria arrived next, and they hit it off. Surprised me a bit, since she wasn't exactly an outdoorsy type of person. Emma was."

"Also, both were mega peeved about the lack of magical progress."

"Eh, yes. It became a fixation for them. I mean, you arrived on the scene late. You weren't even aware of what was going on, and you compounded your sins by shooting right to the top of the magical food chain. It didn't sit well with them."

Did it ever. "I know."

"Then Eve died, and Dot fell ill, and we were stuck in a limbo."

"Because I refused to pick up the reins until Dot...uh, went ballistic."

"Yes. To make matters worse, you have a familiar. And a grimoire, like the Wytchetts. It was jealousy at work, nothing else."

"Jealousy can turn to spite, and that's what I sensed. A smoldering, festering vengefulness."

"Hmm," Elmsworth said. "Gloria and I weren't best mates, but I knew her as a very self-controlled person. I'm not sure she would allow herself to hate. Anyway, the point is moot since she's dead. Emma has more of a temper on her, though she's not unreasonable. Unless we're talking hexing. There, she has a fixation."

He had hit the sore spot. "This is exactly the point where I get stuck every time. It looks like we're dealing with an accomplished practitioner. Someone a lot more competent than any of us. She must've been trying to confuse us."

Elmsworth pointed a bony finger at me. "To me, the frustration rang true, though I might well be wrong. I guess you have an answer for that as well, eh?"

"Unless I'm barking up the wrong tree, her true motive has nothing to do with magic. I can't tell you what it is, but the whole brouhaha over the lack of training was so overdone, it must've been a smokescreen."

"You need to understand that Emma's life jumped off the rails at some point. She's divorced twice. She has no children, is allergic to pets, and she's too intense for most people. It all adds up. She's lonely. I always thought magic must've become the focal point of her hopes and dreams, her big lifesaver. Given the overall lack of progress, she blew a fuse."

"Loudly, yes."

"Mind you, I'm not trying to excuse her actions. If it's her. At this moment, we can't rule anyone out. That's actually my point. Your theory pivots on our antagonist being a coven member. What if you're wrong?"

That wouldn't surprise me in the slightest. "I might well be. Yet how would this person know so much?"

"Scrying? If she can teleport and take on appearances, she could also spy on us from afar."

The fine hairs on my neck rose and prickled, as if a pair of invisible eyes were trained on me. Or an equally invisible gun.

"Ouch."

"Exactly. We don't know where this person's limits are. Surely they'll have them. For the time being, we can safely assume they're indeed in control of their skylles. Hence my doubts concerning Emma. If she had skylles, she hid them well."

My neck prickled some more. Elmsworth was right. Preoccupied as I'd been with my worries about Chris, I hadn't thought things through fully and cut plenty of corners to pinpoint the culprit. "Maybe she isn't the killer. But she's done something. Why else would Mrs. Mornings say what she did?"

Recently, inspiration hadn't been a big thing with me, but the bright flash of sudden awareness lighting up my cranium felt like it. "Ah. The letters."

Elmsworth quirked a bushy brow, sandy-red mixed in with gray to match his mustache and sparse hair.

"Some letters were hand-delivered. Others were mailed from France. Unlike the missive I received in the post, the ones delivered by hand gave me no weird sensations whatsoever."

"Yes, and?"

"What if the witch isn't a coven member but has recruited helpers from our midst? If they, rather than the witch, handled the local letters, it would explain why Daisy's and Jen's missives gave off no nasty sensations."

"Ah, helpers as in Emma?"

"And Gloria. Though why kill one of them?" I jumped up. "We're going round in circles. I must talk to Emma. Now."

Elmsworth rose as well. "The cops might've beaten us to the post."

"If Emma is our killer, she'll be in hiding. Otherwise, she'll simply give Sarah a hard time. Should I spot any officers lurking in the bushes, I'll turn the other way, but we've got to get there first."

"Indeed. I'll come with you. Emma listens to me."

"Good to know she listens to someone."

—

Avebury's High Street teemed with the typical mix of tourists waiting for the manor to open and hikers stocking up on carbs and fluid at the same grocery that also drew in the families. To my tired eyes, there seemed to be fewer Pagans than usual, but as if to compensate, I spotted two aging punks and a Buddhist nun decked out in saffron hues. At least two stag and hen parties were in town, wearing themed T-shirts, pink piglets for the men, skulls with Viking helmets for the women. Between the kids and the party folk, the decibels were getting rather painful.

Elmsworth fiddled with his mustache. "Come lunchtime, and they'll be as drunk as skunks."

"The kids?"

"Har, har."

"Don't worry, our fearless pub keeper won't let them."

Elmsworth pointed at a grumpy-looking Husky, mushing a chariot bike trailer filled with beer bottles. "Guess what, it's BYO time."

"Summertime and the living is easy?"

"Until the next rain shower." He eyed the sky where clouds sailed over skies so blue it was as if they would never, ever dare to dump moisture on us.

Past the cemetery, High Street ran out of quaint cottages, and somewhat more functional accommodation took over. Farther on, Broad Street sheltered the posh places, but this part of town was a sort of in-between spot, stuck on the edge of being acceptable without quite making it to okay.

Emma Bingham rented a first-floor apartment in a building which once started its life as a cottage. At some point, someone had the bright idea of plastering over the stones, putting on an extension, and painting the lot an odd shade of cream, a bit like rice pudding gone bad. The green streaks from the drainpipes didn't improve the visuals one bit.

Cottages around here weren't that big, and dividing them

into two flats wouldn't help with the size, extension notwithstanding. Nor would the rent be in tune, not with the housing market being the way it was.

These thoughts and others in a similar vein tumbled through my head as the Colonel and I followed the weedy path to a scratched entrance door that wasn't quite dingy yet but working hard on getting there.

"No coppers, hooray," Elmsworth boomed.

"Psst, not so loud." My voice sounded a tad too breathless for my taste. To be honest, this amateur sleuth business wasn't as easy as my favorite mysteries made it out to be.

The Colonel pushed the doorbell of the upper flat.

Nothing much happened, so he mashed on the button again.

When a window slammed open, I jumped what must have been a mile high, my heartbeat catapulting into the stratosphere.

A young man wearing the obligatory baseball cap backward stuck his upper body out of the window, releasing an odd odor where stale beer and unwashed clothes were the more pleasant components.

"Oi, what's up with the old Yank broad today? First it's the pigs, now you lot."

I took that to mean Sarah's cohorts had visited already.

"Good morning to you too," said Elmsworth, his pronunciation at its most precise. "Did you see Emma by any chance?"

"Who?" Ball cap's grin turned nasty, showing a row of crooked and rather sharp teeth. "Oh, you mean the old she-hippo. Nah, haven't heard from the broad for a while."

Elmsworth and I swapped a meaningful glance. At least, I think we shared the nasty thought that had taken a grip on my mind. Emma might not have been around because she was busy killing people and teleporting between the UK and the continent.

"Do you remember when you last saw her?"

The young man scratched his chest, clad in a vest splattered with the stains of his last meals. He pointed a dirty finger at us. "I get it. Plainclothes, that's wotcher are, eh? You won't fool me. But I'll tell you once and for all I don't care for the bitch. Clomp-clomping around in her pad all day long. Tap-dancer in clogs. That's what she is. Do you lot ever react to complaints? No-oh, you don't. You won't get a thing from me." He slammed the window shut.

"This didn't go so well," Elmsworth observed.

"Poor Emma, having to share a house with a yob like that. Eh, why did he call her a Yank?"

"Her first husband was American. She worked in the States for, oh, I think, almost a decade. Whenever she wanted to annoy people, she put on a strong Texas accent."

"She never did with me."

"I think she might have quite a bit of respect for you. Anyway, after the divorce, she returned home and never found her professional feet again. In her last job, she was acting as the sales rep for a pharmaceutical company somewhere up north, but they closed, and since then it's been impossible to land decent employment. At least her parents left her a small sum of money, just enough to live on."

I stared at the house in front of us, shabby and downtrodden, with the neighbor from hell on the ground floor. Some life.

Would an accomplished witch live in such a dump? "I can understand why she would be keen to discover her skylles."

"Yes. She doesn't have much else. Anyway, with the cops having been and gone, we could have ourselves a quick sniff."

"What about keys?"

Elmsworth grinned and tapped his nose. It was slightly crooked, making me wonder whether he had played rugby at one point.

"I know where it is." He disappeared around the side of the house, and when he returned a few minutes later, a key dangled from his finger.

"Ah."

"Taped to the drainpipe, like she said. Sunnyboy's window on that side is stuck, so he couldn't see what she was up to. Nor is he the gardening or fixer type. No big risk of him finding it." Elmsworth dragged two pairs of plastic gloves from his pockets. "Here, you better put these on. From what I've seen of your friend, she'll know you've never been in Emma's flat and will have your guts for garters if she spots your dabs anywhere."

"She'll be trying for a search warrant as we speak. Thanks." I slotted my fingers into the slick gloves.

The key must have been cut at one of these express locksmiths since the lock put up a fight, but eventually we were in.

—

Emma Bingham's flat came as a surprise. Sure, an interior designer might have considered the furniture to be too heavy for the small rooms, but every piece was polished to the nth degree. The abstract watercolors on the walls glowed with the warm reds, pinks, and ochers of Canyonlands, as if the hot breath of the desert were blowing straight into my face.

"Those are great," I said.

"Aren't they just? A shame she stopped painting. She's incredibly talented."

"Emma created these? Wow." The bits I didn't know about my fellow coven members filled a book. What sort of top dog did that make me?

Inch by inch, the Colonel pivoted on the heels of his solid walking shoes. "Remind me. What are we looking for exactly?"

"That's the problem. I couldn't tell you. Petals of the type you received in the envelope would be one thing. A grimoire, maybe? The magical scraps her family inherited weren't worth writing home about."

"That's hardly Emma's fault."

"Colonel, I wasn't blaming her. All it means is that what she handed in wasn't big news. But she might not have handed everything in. Otherwise, we're looking for anything that proves she might've been in France recently. Though I suspect with all this teleporting, our antagonist won't have the time to collect physical evidence like bar bills."

"Hmm. I find it hard to believe magic could ever be an ideal means of transportation. At the very least, her constant flitting around would cause one heck of a backlash, correct? When I used my skylles, I wasn't fit for human company for quite a while afterward." He pulled a face.

"Listen, I don't have the faintest clue how this sort of advanced magic works. That's what bothers me most. I can only hope either Jen or myself finds some useful information in our grimoires."

Elmsworth chuckled. "Happy searching."

His words, uttered in his calm, deep voice, worked wonders for my inner worrywart, and with renewed confidence I searched first the tiny kitchen and then a cramped shower room with mud brown and olive tiles straight from the seventies. Everything, including the grouting, was spotless.

No grimoire, though. No petals, nothing more French than a half-empty flagon of Chanel No: 5 in the medicine cabinet. The kitchen cupboards were equally bare. Tins of sardines, eggs, baked beans, a pack of brown toast, and a jar of yeast spread spoke of a need to stock up on calories on the cheap.

Our coven meetings must be a source of much-needed nourishment in what struck me as being a tough existence, lived hidden behind the last shreds of respectability. With a pang, I remembered the carrots and beef casserole Mrs. Bingham had left for me when Aunt Eve died. It must have cost her dearly, and what did I do? Pour it away.

Why didn't she say something? I'd help her out anytime. But she'd never accept it, would she?

Ask discreetly. That's assuming she isn't the killer, of

course.

With a tight throat, I continued my search.

"Myrtle?" The Colonel hollered from the bedroom. "Can you come? Found something rather odd."

I chased into the bedroom, where Elmsworth was standing on a chair and poking a pen at something stashed atop the curtain rail. He clambered off the chair, the object now dangling from the pencil. He shook his head. "She claimed she lost the darn thing. Maybe she forgot, but why would she store it up here in the first place?"

"The darn thing" was a dark olive bucket hat like a fisher would wear.

Chapter Sixteen

The flat had revealed its one secret. At least we found nothing else of interest. With the bucket hat restored to its hiding place for the cops to find, the Colonel promised to put the word out on Emma, and I drove home, my guts all knotted up with impatience. Everything was taking much too long. I was unlocking the door when Jenna's dented transport rolled up.

High-pitched voices yelled, "Auntie Myrtle, Auntie Myrtle."

Jenna had brought the twins.

She cranked open her door and hopped outside. "Sorry, Myr. There's a stomach bug making the rounds at kindergarten, and I wanted the boys out before they catch it."

From my experience, bugs in kindergartens spread at lightning speed, and the kids had most likely caught it already. But what was a little virus among friends?

"Come on in."

"Love to. Would you like some cocoa? I pulled them out before their brekkie-break. But I needed to see you."

My heartbeat ratcheted up. Good news or bad news? "How

about some ice cream?"

Big mistake. Like a red-haired and freckled double twister, Robbie and Johnnie charged into the corridor, shouting "ice cream, ice cream" at the top of their powerful voices. Once seated at the kitchen table, steaming mugs and vanilla gelato within reach, the boys quieted into a picture of domestic bliss.

My brain, always wired in a weird way, projected a clock. I could even hear it ticking. That was unfair. Lots of people had kids in their thirties or even forties. Nothing was ticking for me, thank you very much, so no need to give me grief.

My brain yanked in an image of a tall, swarthy man sitting in a cell. His absence was one minor obstacle to parenthood, even if I—and my partner—were thinking of such things.

Other than chasing after the witch, there's nothing more I can do for you, Chris.

Something dark clawed at my heart.

Jenna, regarding her offspring with a fond smile, didn't seem to notice my glum mood. Then she looked up. "Sorry, Myr, but the laundry list hasn't much to say about changing location by magical means or glamors."

"Nothing much meaning what exactly?"

"You know our grimoire. It's a treatise on magic, not a practical guide like yours. It claims such stunts are reserved for the truly gifted. Adding to that, we have our usual warnings concerning the first directive."

"It's never evil we serve," we both intoned in a lugubrious tone. Suddenly, the sunny kitchen appeared to be filled with ghosts of the past, restless and vindictive.

"Mum, what's wrong?" Robbie licked his lips.

"Sorry, sweetheart. Just your mum and Auntie Myrtle being silly. You better eat your ice cream before it melts."

He dug in.

"Apropos evil. Was there any reference to side effects or limitations? The Colonel thinks it's impossible for our witch to bounce around at her leisure and look like whoever she wants without triggering some fallout."

"Nothing on that, though I tend to agree with him. This comes at a cost, even if it doesn't seem to stop her. Your itchy hex must be the only fly in her ointment."

"Fun fact—Petty thinks I'm every bit as good as our witch. Heh, she loves me, so she would say things like this."

Jen rolled her eyes. "Sweet Earth, aren't you being a tad defeatist here? What does the recipe book say on the matter?"

"I haven't got round to checking the empty pages yet. For sure, Great Aunt Petunia wasn't into glamors and teleporting. Not that she'd use the exact term, but I found nothing on the subject."

"Fancy giving it a try? I'm dying to watch you in action."

"Fine, but what about the kids?"

Two faces covered in melted ice cream grinned at me. "More?" the twins asked as one.

—

Between the balls and tennis rackets Jenna always kept in her car, chalk to draw football pitches on the pavement, and a determined instruction not to climb any trees, Robbie and Johnnie appeared content to leave Jenna and me alone for a little while. To my everlasting amazement, Tiddles extricated herself from her cozy nest and stalked after the boys. Squeals of delight confirmed cat and kids had connected, and so far there weren't any casualties.

Jenna curled up on the sofa, hugging a pillow. "Let the show begin."

Petty had nothing better to do than land her pot next to my friend. Next, she'd ask for some fertilizer balls in lieu of popcorn.

"If you insist. Here comes *Much Ado About Nothing* presented by Myrtle Coldron." I passed under the archway into the den with the recipe book mushed to my chest.

"Where are you going?" Jen hollered after me.

Petty rustled and swished her leaves in silent laughter.

This wasn't good. I was concentrating on everything but calling my skylles. "Movement is one ingredient in the special Coldron magic," I shouted over my shoulder. "I'm totally hyped up on emotions, which is another check. And there's been a murder."

Urgh. That topic was a sore spot with Jenna.

"I'm glad there are no invisible pages in the laundry list," was all she said.

"We better discuss this later."

"Oops, sorry. I'll shut my trap."

Rats. Hexing was difficult enough on its own. With an audience around, this would be well-nigh impossible. I'd never been a quitter, so I paced some more.

Was the book vibrating into my rib cage? Could it also be warming my chest, like my froggy green hot water bottle did?

It would be a real hoot if this worked at first try.

Returned to the living room, I placed the recipe book on the coffee table, its panels closed.

"Right-oh. Recipe book, I need information on teleporting...changing location using the skylles and, uh..."

"Taking on appearances. Glamor," Jen hissed.

"Changing location and glamor, sure."

The front panel of the Coldron grimoire quivered as if to acknowledge my request and flew open, the pages rushing past as if this were a flip-book.

"Whoa." Jen and Petty rose from the sofa to peek over my shoulder. Petty's lemon happy scent filled my nose, and a spark brushed past my cheek, lighter than a cotton ball.

Whoa didn't half describe it.

With a dry rustle, the book stopped reading itself. The parchment flattened, and text flowed in from the sides, sepia-toned at first until they turned a washed-out blue and stayed that way. I had hit the jackpot and triggered the empty pages into revealing their secret without too much of an effort. And without Daisy either. She wouldn't exactly whoop with joy, but her sensitivities were for later.

Petty whooshed past me and landed next to the phone on the table. Then she fired off pink and white sparks.

"What's she doing?"

"Wanting me to take a photo, in case the text vanishes, which it did once." I pointed the camera at the text. Not particularly long, it featured some illustrations.

"Okay to sneak a peek?" Jenna asked.

"Sure."

Jenna bent over the page and was still reading when the phone twitched in my hand and warbled opera tunes into the room. The screen displayed the country code for France, which made my heart jump into my throat.

"Myrtle Coldron speaking?" *The lawyer, let it be Chris's lawyer.*

"Yes, 'ello. It's Ella here. You and your cousin visited yesterday."

Had it only been yesterday? I swallowed, willing my heart to stop fluttering. "Of course. How are you, Madame?"

Jenna looked up from the text. She opened and closed her hand, mimicking a person talking.

I switched the call to speaker mode.

"Fine, merci. I remembered something."

Oh? "It's kind of you to call."

"Avec plaisir. Ecoutez. Someone just visited my shop, and they spoke like this difficult customer we discussed. This woman who bought the potpourri, vous savez?"

My heart fluttered faster. "Carry on. I'm listening."

"Americaine. J'en suis sûr."

American. Wild thoughts crashed and rolled through my mind. The Colonel mentioned that Mrs. Bingham had lived in the States and could fake the accent. Perhaps none of this counted as incriminating evidence in Sarah's world. To me, however, it sounded pretty much like a final nail going into Mrs.

Bingham's coffin, even more so considering the bucket hat she claimed she had lost.

"This is interesting."

"Does it 'elp you?"

"Oh, yes. Merci mille fois, Madame."

"Ah, it is a pleasure. Say hello to your sister for me." She pronounced it "sistair."

"I will. On my side, I wish you many happy customers."

Ella laughed. "Merci." She cut the call.

"I gather this was important information," Jenna said.

"This was the person who runs the shop in Carnac where the witch bought the potpourri. She's just remembered the buyer had an American accent. I didn't know Mrs. Bingham lived in the US, did you?"

"Nope, though she drawled when bored. Not like Greg, but there was something...oh dear, does that mean *Emma*'s our mysterious witchy killer?"

"Let's say it doesn't look good, since Elmsworth and I spent quality time in her flat and found something related to the other killing." I recapped our discovery of the bucket hat.

Jenna did a great deer-in-the-headlights impression. "But she's such a total wuss when it comes to magic."

"That must be what she wants us to believe. Speaking of magic, what does this text say—"

Once more, the phone twitched and buzzed in my hand, this time playing the theme from Star Wars.

"It's your brother. Hello, Marty."

"A few minutes ago, Gloria Mornings was in our living room, trying to steal the laundry list."

Gloria Mornings was dead. A deafening silence fell.

Jenna cleared her throat. "It's the witch. Good, she doesn't know I keep the box in my bedroom."

"Few do," Marty said.

I'd known about the temperature-controlled box, but not where it lived.

"How did you know what she was after?" Her eyes widened. "Sweet Earth, did she spot you?"

"Nope. I was out in the garden. Heard furtive noises and crept up to the window. She was babbling some sort of mantra, mentioning the laundry list."

"Marty, are you sure it was Mrs. Mornings and not Mrs. Bingham?"

"They look quite different. Plus, Mrs. Mornings is dead. That's how I knew I had a problem."

"Did she have an American accent?"

"Didn't pay attention."

Shame about that.

"What she did was scratch herself the whole time. When her face sort of wobbled and slipped, she dashed outside."

Wow, my impromptu hex seemed to be quite durable.

"I've now locked both entrances."

"Forget it. Apparently, she can hex locks," Jenna said, her voice grimmer than a November morning. "I'll have to hide the grimoire in the cellar."

The cellar under the farm wasn't the ideal environment for ancient tomes. Somehow bigger than the farmhouse above, its cobwebby brickwork echoed with ancient menace—what was Jenna saying?

"...hide the plaques down there as well. And our mirror."

"Would make sense," Marty said. "Sorry, ladies, gotta dash. Delivery's due. If you need me, just holler."

The call finished. I turned to Jenna. "Are you sure your basement's the right place for the laundry list?"

A shadow flitted over Jenna's elfin face, as if an invisible cloud was passing. "Our cellar is a bit special."

That was one way of describing it. "And here I am, thinking I was suffering from an over-active imagination."

Jenna raised a brow.

"Your cellar appears much bigger than it can be. As if space and time get distorted down there."

Jenna's face stiffened. "Gran swore there must be a hex on it. What you mentioned about the size of the place is actually a pretty accurate description of what's going on. Gives me the creeps." She shuddered.

You and me both. "How old is it? And how old is the farm? For me, the two don't match."

"They don't," Jenna said. "The cellar predates even the first farm, which was built in 1790."

Somehow, the words didn't want to be spoken, but I squeezed them out. "Could your basement date back to the Earth Wardens?"

"That's what Gran always thought," Jenna said. "No matter what, it's a great hiding place for things you don't want found."

"We'll need the mirror again next summer for the anti-curse ritual. Even if we don't know the exact purpose of the plaques, I'd rather not lose either."

This time, a faint smile tugged at Jenna's lips. "I don't know what purpose the cellar served, but for us it seems to—well, like isn't the right word."

"Uh-huh. Are you sure that place won't absorb our stuff into its brickwork?"

Petty rapped on the table twice.

"This means no," Jenna said. "She's right. Whenever I want our cellar to hide something for me, it will do so. Until now, I always got my stuff back without fail. "Believe me, for the moment, our cellar is the coven's best bet. Oh, will your recipe book be safe? It's hexed, isn't it?"

"Yes. It hides from the Simpkins sisters, so I'm reasonably sure the witch can't get at it."

Petty rapped out a yes, and I breathed a sigh of relief.

My respite was short-lived. How would the magical plaques and the mirror we needed for the coven's survival be safer in a spooky cellar than in a high-tech bank vault? But this was about more than trust—this was about changing my outlook on life, on things like facts and reason, things I'd taken for granted my entire life. Either I trusted Jenna's take on magic or I didn't. It was as simple as that.

I heaved a deep breath. "If you say your cellar is a safe place, then this is where the plaques and the mirror should go. I'll fetch them later today."

Smiling, Jenna squeezed me in a teddy bear hug. If there was such a thing. "Thanks for believing in me."

"The Misfits are short on magical whiz kids, but you're definitely one. Which brings me back to our text." I raised my chin at the recipe book, still open on the table.

"Haven't finished yet, but there's a blood spell involved."

Of course. It would be something rotten like that. "Bad news?"

"Not necessarily. Just super difficult. Almost impossible to pull off. Also, you'd need blood from a living creature, hence the name."

"Whose blood? And how much?" That had come out a lot squeakier than intended.

Jenna furrowed her brows. "It depends. Actually, something bothers me here. If you mail me the text, I'll have another go at it. I'll also cross-check against the laundry list. I think I know where to look."

There was a whole dung heap of bits bothering me here, so that made two of us.

"I'll get moving on the cellar front." She dug in her pockets and handed me the key to the bank vault. "You better take this back. When does the bank close?"

I glanced at my watch. "In about two hours. Means I too need to get moving. Uh, shouldn't we involve the Ragworts and Elmsworth? And maybe Daisy and Mel? We don't have time for a full coven meeting, but there's safety in numbers."

"Great idea. I'll ring them once I've ferried the rascals back home. See you tonight." Jenna gave me another quick squeeze before dashing from the room in search of the twins.

Chapter Seventeen

A lone with Petty, I weighed the phone in my hand. Should I tip Sarah off about the American accent?

You're a witch. Use magic to find the beast and shut her up.

No go. Between my headache and general tiredness, scouring the recipe book for a tracker spell would knock the stuffing out of me. That was assuming our grouchy grimoire would accommodate me twice in one day.

Anger flared in my chest. The whole situation was impossible, but anger would get me nowhere. Embracing magic didn't imply turning my back on things like logic, and logic decreed Sarah was an ally, not an enemy. That being the case, she'd need to hear about the American accent.

Basta, pasta.

At least the accent was a proper clue, and to get it, I committed no crimes. My text message sent, I felt relieved, even mildly virtuous. A second later, I stopped doing that. I still had this morning's visit to Mrs. Bingham's digs on my mind.

A ping announced an incoming message.

Sarah.

"Thanks. Every bit helps. We've searched Mrs. Bingham's flat, and while details are off limits, you might like to know she's a person of interest. I'll contact my French colleague. Looks like you're right, and there's a connection. It might help your Chris."

She'd added a grinning emoji at the end of her text.

Phew. For the moment, life was looking up. Not a biggie when disaster had this tendency of piling up and threatening to crash down on me, but as Sarah said, every bit helped.

Should I ask Jenna if she'd made sense of the recipe book's intel? But no, she'd call me as soon as she could.

Ping. As if to reward me for showing common sense, a new message rushed in.

Kerluac, the lawyer, had talked to the magistrate, who seemed outraged over the copper's incompetence. It appeared the wheels of French justice were grinding into motion. He'd also had a word with Chris, who said to share an enormous hug.

At that point, I teared up but still briefed Kerluac on the Bingham developments. Just in case the French side of law enforcement pulled more tricks from their sweaty caps.

Every bit helped.

Another glance at my watch spurted adrenaline into my system. Now I needed to swing my hoofies. Of course, my gaze then stumbled over the recipe book, still lying open on the coffee table, still displaying the mystery text.

Behind it, Petty had parked her pot, sparking like a wildfire.

"Oh, drat. Now's not the time for more hexing."

Yes, Petty rapped.

"No can do. Jenna's on that text and once I feel better, I'll search for a tracer spell." Petty rapped once, but didn't cease her sparking.

"Sweets, I must get to the bank before it closes."

My familiar drooped her leaves, and I jumped on the slide that led straight to the guilt pit. "You know what? I'll give Daisy a call. She can take you along when we explore the cellar. How does this sound?"

In a weak rustle, Petty shifted her flowers. The sparks winked out.

Enthusiasm looked different.

A thought sailed in. Was I being insensitive? After all, the cellar had been the place where I killed my poor familiar when I rejected my magic.

"If you don't wish to go back down there, I understand. I simply thought you might like to come. Just tell me yes or no, and we take it from there."

She rapped out a quiet yes.

"Fine, then I'll buzz Daisy and put her in the picture."

Once again, my gaze hitched at the grimoire. Since I didn't dare trigger my cousin's insecurities about her skylles, I'd better not tell her I coaxed the old tome into action on my own.

Message to Daisy delivered. I climbed upstairs. My bed beckoned, but my leaden limbs and aching head would have to tough it out for a little while longer. Refreshed by a quick shower, I traipsed back down and broke the Wiltshire speed record to get to the bank.

The clerk took his password checks seriously, which popped all my doubts back up. Was I sure about this? Was I doing the right thing?

No one knew, so I returned to my car carrying the strongbox with two Neolithic plaques, the so-called keys of the witches, and the rusty mirror we'd lifted from Mary Anne's grave and used to fend off the curse.

All the way back to Avebury, I kept checking my rearview mirror, but no cars were tailing me. Perhaps I was too tired to notice. Perhaps Fate was smiling on me.

A fat raindrop burst on my windshield the moment I steered through the s-curve leading through the village. My environment had turned a shade of menacing green, and even with the steady purr of the motor, the stillness outside registered. Not a leaf trembled on the trees. A funereal cloud mass was rolling in from behind me, towering high and displaying the sort of flattish top that spoke of seriously severe weather.

Dread closing my throat, I stepped on the accelerator and made it to the farm entrance the same moment the weather broke.

Fate had a funny way of getting its own back.

—

Rain sheeted down the windows, blurring my environment into meaningless blotches. Lightning streaked across the sky faster than the thunder could catch up. The result was a cacophonous banging and rumbling, carried on a storm buffeting even my solid van. The world had gone from green to pitch-black, the indifferent glare thrown by the van's headlights notwithstanding. Only a few meters separated me from the entrance to Wytchett Farm, but even one inch would've been too much.

The display on my phone lit up. Incoming call.

If it now plays "Riders on the Storm," I swear I'll throw it out of the window.

It didn't. It played "Twinkle, Twinkle Little Star."

Jenna, the display read.

"Hello?"

"I can see you sitting outside. The rest of the gang's already here, including your cousin. And Petty."

"Good for them. I'm not getting out of here."

"Of course not. What I wanted to tell you—there'll be a break in a few minutes. After that, it'll be worse."

"Break meaning what exactly?"

"Well, the first front is moving out, but there's another one behind it. We have a tornado warning. Witching weather, the Met Office called it."

"Oh, joy."

"You don't want to be out in this. As soon as the rain eases off, you better dash. I'll open the front door for you."

"Okay." I pocketed my phone and grabbed the strongbox, ready to slam from my transport.

Was the fury of the wind abating? Was the manic blaze above me dimming?

A bright flash flared, immediately followed by a sharp crack that lashed my ears.

The primal fear of all creatures caught by the elements shot up my throat, and my scream ricocheted in my van. Torrents of rainwater—or were those hailstones?—rattled on my roof.

Witching weather? What if the witch was behind this and she lurked outside, hiding behind the forces of nature she'd unleashed?

Suddenly, the night-like gloom grew eyes. At least in my mind, it did.

Another flash, two of them, and more crackling that sounded as if someone up there was folding the biggest aluminum foil known to humankind, but this time the van didn't rock. The rumble died away, a drum roll fading.

A second later, the rattle on the roof changed into a hissing noise. Where Niagara Falls had been flooding the car, simple rain now pelted the windscreen. A lot of it, yes, but the weather no longer seemed fixated on washing the van complete with its passenger into the Irish Sea.

Only now the scent of wet soil and growing things register. The carpet was vibrating, and there was a greenness creeping into my vision. Pink rose petals fluttered through the air like frightened moths.

I wasn't alone in the storm. My skylles were with me.

The panic from a few seconds ago died down and, in its stead, a roaring fury spread through my system. Why couldn't I teleport like our witchy antagonist? She wouldn't be sitting in her car like a lost whatnot. No, she'd rip holes into space and slip through them onto the farm, like the supernatural beast she was.

A vision of an open door formed in my mind, but it washed away. I would get soaked. Not she.

So. Not. Fair.

More crackling of aluminum foil, followed by a hissing noise. There was another flash, this one orange, not white and bright. A whirling sensation swept me off my feet, and next I was dropping into nothingness as if the driver's seat of my van had morphed into the Tower of Terror. Fairground rides and I were polar opposites; the few times I didn't want to be a party pooper and squeezed into a rickety seat with only a paltry bar between me and instant death, I paid dearly for my idiocy.

This experience was worse.

On the flip side, it didn't take long. One minute I was plunging into nothingness, shrieking; the next I stumbled through Jen's corridor, the strongbox still clutched to my chest. Somehow, a wall jumped at me, blocking my progress more effectively than an uppercut into my rib cage. Okay, walls didn't jump, but that's how it felt.

"Oof." I slid down the useful but unyielding obstacle.

"Oopsie, Myr, where did you come from?"

Overlaid with a ringing in my ears, the voice sounded a lot like Jenna. Liver-colored splotches and fuzzy sparks, which mimicked Petty's fusillades, drifted through my vision.

"Door?" I croaked.

"I never opened it. I was ready to, but you were sitting out there. Until you..." Her eyes widened. "You fell from the ceiling. Through the ceiling, actually."

"Huh?"

"You must've teleported. I was right. Oh, Myr, I was right." She raced up and squeezed me with another of her teddy bear hugs.

What might be considered an achievement of epic proportions made me feel exactly nothing. Yay, I had teleported, so what?

Was my stupor a result of too much running around after

not enough sleep, or was it the hexing? Possibly both. At some point, the human body shuts parts of it down to protect itself, and I could have sworn mine was doing just that. Unfortunately, there was no time for navel-gazing, let alone some downtime to catch my breath.

We needed to hide the plaques.

When I put thoughts into words, Jen responded. "If you insist. But you look awfully pale. And I'd say you have a few more gray hairs." She reached out and gently tucked a strawberry strand behind my ear.

"What do you expect? Most of the time I go without sleep. Should I snooze off, Chris is sure to haunt my dreams. When I'm not having visions of witchy burglars. I want this over and done with."

—

Using the chilly handrail for support, I followed Jenna down the worn brick steps into the depths skulking beneath Wytchett Farm. Here and there, the mildewy gloom was broken by bulbs dangling from the ceiling, which as a source of light left a lot to be desired. As cold as the handrail might have been, it was nothing against the arctic chill that made Jen's and my breath billow in white, as if we were some weird sort of magic dragons.

Jen slipped around the shelf facing the nearest archway, hosting preserves, jams, and empty apple crates. A few archways farther on huddled a group of people equipped with flashlights, which Jenna joined, talking at the top of her voice about my latest stunt.

Dread tearing at my stomach, I searched for my cousin and found her standing close to Damian, her expression shifting to sullen.

Who was I to blame her when magic played favorites and broke all the rules? One teensy boost would mean the world to her, but it didn't happen, while I steamed from strength to strength without ever wanting to. Lore allowed one strong witch per family, and it appeared I was it.

Oh, joy.

A starburst of sparks lit up the darkness, revealing Petty's presence. The next instant, she zipped across the underground airspace to hover in front of my face, bobbing up and down like a gull on a swell. She came so close her blossoms tickled my cheeks. Never had her lemon scent been more intense. By now I knew her well enough to know what she must be thinking.

"Yes, fine, you were right. Seems like I might have the odd trick up my sleeve. Just don't ask me for a repeat performance. I've no clue how this happened."

"Doesn't surprise me in the slightest," Rosie said. "You're special."

Mel, unflappable as always, sailed across, and for the third time today I was smothered in a gentle embrace. "Girl, I'm proud of you."

Sudden heat burning in my cheeks, I returned the squeeze and then wriggled from her arms. "Talk to me once I've processed this crap."

Daisy hadn't budged, and when I sought her gaze, she looked away, radiating frustration. I couldn't change what happened, but maybe I could stop the hurt.

I stepped up. "Daise, I'm sorry. I swear I don't do this deliberately."

Her sigh must have traveled up from an abyss. "I know. It's just..." She turned her head aside.

Our shared adventures made my next move easy. I placed the strongbox on the ground, pulled my cousin to my chest, and held on to Daisy's trembling warmth.

To my everlasting relief, she slipped her arms around my shoulders and hugged me back. She also sniffed my shoulder. "You smell like a forest."

"How does a forest smell?"

Daisy's giggle teased my ear, easing the guilt weighing on my chest. She'd get over it, and we'd still be friends. "Woodsy, spicy, and somehow green." She took a step back and gave me a once-over. "You look tired."

Tell me something new. "Isn't it odd my magic manifests as rose petals when it smells of the woods?"

"The skylles are born from the soil," Jenna said from behind. "But each witch makes them manifest differently."

Born from the soil? I couldn't convince my fuzzy mind to drill into the concept, and the others didn't help with their excited babble. Plus, it was too bloody cold to think.

"Shall we?" Marty pointed at the strongbox.

"Sure," I said. "How far in do you think we have to go?"

"Not too far. Welcome, folks, your guided cellar tour starts this second."

No one laughed.

Daisy's warmth already forgotten, shiver upon shiver frosted my spine. This place had it in for me; I knew it. Not only did it echo with crappy personal memories, the solid brickwork marching into the distance seemed to whisper of other times, other people, who once walked the Earth, breathing, dreaming, worrying like I did now, their stories forever lost.

A gentle nudge on my hip broke my musings. A spark sailed past my eyes.

"Thanks, Petty." I needed to watch myself. Recently, I slipped into weirdo mode at the slightest provocation.

Beyond the part of the cellar Marty and Jen used for storage, we trudged through the echoing darkness. Illuminated by the wobbling beams of our flashlights, archway after archway peeled from the gloom, only to disappear again. As far as I could work out, we were going in a straight line, if straight meant not running into the pillars between the arches. The ground under our feet was gently sloping downward, but I might have been mistaken about that. With the erratic patches of light skittering from bricks to packed earth and back up again, my sense of direction had given up its ghost.

"Almost there." Marty stumbled over a rusty scythe leaning

against a pillar. It clattered to the ground. "Turn right, please. The green thing you see over there is a window."

"What green thing?" Damian was blinking myopically over the rims of his thick glasses.

"This, dear." Rosie pointed at a smudge in the distance that, with a bit of goodwill, might be called green, though sludge-khaki was a much better adjective.

"It's a window," Jenna said. "At least we think it is. Don't ask me where it is, because we don't know."

"Eh, these days it's easy to pin down locations," the Colonel said.

"Nope," Marty said. "Be my guest and try. We did, more than once, but we never found it. In terms of distance, we reckon we're not even one hundred meters from the entrance, but distances down here are a real bugger, and GPS is a total no go."

For me, half a lifetime appeared to have passed, and directions didn't count anymore. But Petty was still by my side, floating unconcerned as if enjoying a walk in the park, so we were most likely okay.

"Brr," Daisy said, wrapping her arms around herself. "You sure you can find this place again?"

Jenna snickered. "No matter where Marty and myself went, we always ended up at the window."

"Used to play hide and seek down here as kids. When I hid near the window, people never found me. It was the same for Jen. Our Gran didn't know about it either. Not until we showed her, that is. Then she knew where to look for us when we'd gone into hiding."

"That's why we figured if we take you here, you'll be able to come back," Jenna said.

"Oh," Daisy said. "Sweet of you."

She was right. It was generous of the Wytchetts to share their secret, and we all muttered our thanks, while shuffling faster because of the cold.

The greenish smear turned out to be an almond-shaped opening high in the wall, covered by glass so dirty and fogged up by cobwebs so thick, one couldn't see a thing. The green, though, was most likely moss or algae.

A pile of rotting sacks which gave off the fetid odor of moldering potatoes lay at the foot of the damp wall underneath the window. Next to the pile stood a barrel, some broken chairs, and ancient farm-implements, as rusty as the scythe. As a hiding place, it would do.

Not wanting to touch the icky fabric of the sacks, I lowered the box onto the jumbled remains of the chairs. Despite trying hard not to touch anything, the back of my hand brushed against the wood, slimy with moisture.

"Blech." I wiped my hand on my jeans. "Are we done?"

"Guess so," Marty said. "Nothing more to see."

"Jolly good. Now, we need a strategy on how to deal with the enemy," Elmsworth said.

Damian pointed the beam of his torchlight at the Colonel's face, and he blinked like a startled bat.

"Sorry, mate." Damian's beam shifted, slicing the darkness into ribbons.

"That's why I wanted you here," I said. "Can we please continue in a less spooky environment?"

Jenna was already on the move. "I've prepared a snack. Thought we might all need it. Let's vamoose."

Vamoose we did, and when I looked back at the window, the chairs hiding the strongbox had already melted into the shadows.

Chapter Eighteen

Despite drinking copious amounts of tea and coffee in Jen's kitchen or enjoying Marty's barbecues in the apple orchard, I'd never once visited the living room at Wytchett Farm. Teak sideboards, console tables, and cupboards lined the walls, where fringed textile lamps dating back to the Fifties spread their soft glow into the cozy chaos. Three settees covered in patchwork quilts and throw cushions of mixed design and heritage were made for slumping, and no one needed a second invitation, least of all me. As I lay sprawled over my cushion, I blinked at walls exhibiting a flurry of the twins' masterpieces while, deep inside of me, nerves wound tighter than a bedspring dared to uncoil.

Jen busied herself with a dented chrome samovar, and soon we were holding mugs and slices of lemon tart on mismatched plates. The hot tea and the tangy, fluffy sponge with its crispy icing made me sigh with contentment. Some shuteye, and I might feel normal again.

Jenna cleared her throat. "Myrtle, about you teleporting—"

Ouch. Normal was worlds away. "Not now, please. Let's come up with a plan first and get our heads around the assorted miracles when I'm more *compos mentis.*"

The Colonel, leaning on the bricked-up fireplace, said, "Would you have a whiteboard? I like plans I can put my hands on."

"I'll fetch you something." Marty strode outside.

Damian dabbed at his mouth. "Top priority is catching Mrs. Bingham before the police do." He shot me a meaningful glance. "Since they've been alerted."

Sudden anger gripped my throat and squeezed. With trembling fingers, I placed my mug on a polished tree trunk acting as the side table. Only then did I throw up my hands. "I know, I know. Involving Sarah was a foolish decision on my part, but without her, Chris would be in even deeper trouble. Not to forget, Mrs. B is a killer, and it's a police officer's job to deliver delinquents to justice. Yes, there's a problem with keeping her in jail, and she might rat on the coven. But what else can we do? Lock her into the dungeons?" I pointed at the wooden floor, meaning the cellar below.

"I wasn't pointing a finger but stating a fact," Damian said mildly.

"Sorry." I massaged my aching eyes.

"Jenna said you can jinx a person's skylles and block certain hexes," Daisy suggested. "Once that's sorted, Sarah can have her."

"Easier said than done," I said.

"You teleported," Daisy said as if it explained everything.

Jenna returned, carrying a small blackboard and a box of chalk. "Would this work for you, Colonel?"

Elmsworth's mustache did that twitching thing which betrayed his annoyance, but he said, "It'll do, thanks." He fumbled through the chalk box, sighed, and picked a pink piece of chalk he used to write on the board. "So, Mrs. B tops our suspect list. Aim is to catch the woman before the police do, find a hex to shut her up, and stop her from teleporting. Once that's sorted, we'll hand her over to the coppers. What else?"

"Catch her how?" Mel asked. "She's been pretty apt about slipping through every net. Don't shoot me, but why don't we involve the villagers?" Silence fell on the room. Even the slurping and chewing stopped.

"Crikey," Rosie said. "Are you sure, dear? I mean, first Dot turns out to be a wrong-un, now Mrs. B is headed the same way. We won't win any village popularity contests anytime soon."

"True," Jenna said. "Looks like we've strained neighborly relationships to the max, and that's a fact."

Daisy snapped her fingers. "The Simpkins sisters. Surely, they'd love a chance to stick their oars in."

"Use them as go-betweens?" I asked. "It might work, but I'd rather not force them to choose between us and their friends."

"Try them anyway," the Colonel said. "Without reinforcements, we stand no chance. Oh, and every coven member must be crystal clear on what's at stake and expected of them. Call chain won't work for this. We need a coven meeting."

Groans all around.

"It's always such a *hassle*, isn't it?" Mel summed up my sentiments aptly.

My lethargic brain cells twitched and sparked an idea based on something I'd seen recently. Not something, but someone— Chris in jail. "Video conference?"

The room fell silent.

Elmsworth was the first to clear his throat. "Didn't we discard that idea back in April because of the risks involved?"

"Chris told me with an enterprise plan, the risks of someone hacking in are within what he calls reasonable limits. We didn't go that way because certain people hated the notion." I didn't need to drop names. Mrs. B and Gloria Mornings had been among the first to bulldoze the proposal. Not that they'd been the only ones.

Twenty-one, twenty-two, twenty-three—

"Hmph, I'm no fan of video conferencing," Damian said. "Rosie, dear?"

Bingo. Here we go again.

"I'm not sure. I never tried," Rosie said. "Actually, I don't think I could handle so many people at once."

She could handle tons of people face-to-face just fine.

Marty rolled his eyes. "Do you have a better suggestion?"

Damian slipped his wife a hopeful glance and received a shrug in lieu of a response. He sighed. "Not right now. But I can't say I'm happy with this newfangled stuff. First the database, now video conferences. What comes next? Witches are old-fashioned beings."

"Noted," Marty said. "As soon as we've honed up on telepathy, we review the approach."

Elmsworth wrote "vidcon" on the board, underlining it twice. "So, that's settled. Who's in charge of getting us an account and telling people?"

"I can do it," Daisy piped up.

"Excellent. Myrtle talks to the Simpkinses, and Daisy's on the vidcon, to get everyone on the level. Any more suggestions?"

My phone played "Twinkle, Twinkle Little Star." The same tune twice in a row? Wonders never ceased.

It was the bank. It also was after closing time. My stomach plummeting into my sneakers, I took the call.

"Ms. Coldron? Moleworthy's the name. We're slightly confused here. You showed up earlier and fetched something from the vault, correct?"

The floor under my shoes did a great see-sawing impression, and I somehow managed to put the call on speaker. "Indeed. I proved my identity. You have it all on camera, correct?"

He coughed. "Uh, no. There seems to have been a bit of a glitch, so no recordings. Well, my understanding is an individual arrived straight after you left, waving a written authorization and claiming you had lost the key to your safe, whose contents you needed in a hurry. The woman seems to have been...ehem, let's say, rather forceful. At that time only a junior assistant was available, who was also not aware of your previous visit, and I'm sorry to say she steamrollered him."

"You mean...you let that person into your so-called top-security vault?" My voice rose to a screech.

"Uh, eh, yes? Well...ehem, no harm done, eh? Since you confirmed you retrieved your possessions and, according to the correct procedure as well, everything's hunky-dory, isn't it?" There was a hopeful tone in his voice.

If the thunderstorm had raged outside, I might have teleported to the bank then and there to give him procedures. With the atmosphere having calmed down, I had to restrict myself to teeth-gnashing. "Nothing is either hunky or dory, Mister Holy Moly. Can you at least describe that person?"

"It's Moleworthy," he bleated. "And no, not really. She seemed to be...how can I say? Nondescript? My colleague couldn't remember a thing. In any case—"

The guy was wasting our time. "You effed up, and you know it. You'll hear from my lawyer. Good night."

I cut the call. Looked up at the faces of my friends. "Looks like we scored against Mrs. B."

"Shame we don't know how she looks like," Rosie said.

"That would've been too easy, I guess. I wonder what she'll do next."

—

Telepathy must be a thing since my phone pinged an incoming message, sender unknown, but I knew anyway. "It's her."

All eyes trained on me. I tapped on the message and read it out.

"Feeling clever, do you? You and your so-called coven can rot in hell. I want the magical keys to the henge. Ignatius is to hand over his secret ledger. Tell him he has three days, otherwise he's next on my list. Don't try any tricks in the meantime, or I'll go public. I know how much you hate this. How do you enjoy seeing Loverboy in jail? Blasted Coldrons. It's all your fault. Without Lily's meddling, I wouldn't be stuck in this world. It's high time I joined the others."

As my words petered out, an oppressive silence filled the room until there was no air left to breathe.

Until Daisy jumped up. "Lily? Does she mean Lily Coldron? The Lily who hexed our recipe book back in sixteen hundred something? Is Mrs. B that old? What's this business about the meddling?" More than ever, Daisy's brown eyes reminded me of a spaniel begging for a treat.

My brain refused to cooperate. "How am I supposed to know? It sounds as if our Lily made someone stay behind when the Whites left through the circle, though whether it's Mrs. B or her ancestor, we don't know."

That was when my synapses kicked in, shooting wild images through by brain. Like scared bunnies, they hopped around, looking for a safe hole.

But there was none.

"Oh no. Oh, good heavens, not that."

Daisy's eyes grew bigger. "What's up now?"

"Everything," Elmsworth said. Coming from someone not into drama, the statement packed a punch.

I swallowed, but there didn't seem to be a lot of spittle left in my mouth. "You've come to the same conclusion?"

"That Mrs. B is most likely a White? Well, at this point, I wouldn't call it a conclusion. More like a nasty suspicion. Didn't you call up some text on this subject earlier?"

"I did, though I was looking for intel on teleporting and glamors, not a checklist on how one tells a Red from a White. I thought they'd all left."

Jenna rubbed her nose. "Hmm, your text referred to Reds and glamors."

"Did it?" I fiddled with my phone, called up the copy from the recipe book, and skimmed through the loping script. "You're right. Looks like a step-by-step guide on how to conjure glamors and 'shyfte location by using one's skylles'. Hah, it even says teleporting works best at sunrise, sunset or during foul weather, when the 'aire is heavvy with energye'. Spot on."

Jenna stepped up and peeked over my shoulder. "That wasn't what I meant. Here." She pointed at a piece of text farther down. "This says a Red will suffer from...what's that supposed to mean? Oh, cripes, Lily, your handwriting is a total mess. Difficulties? Yes, she means severe difficulties to 'carrye the ilusion and to shyfte location.'"

There's only so much a grammar nut can take, even when running on empty. "Hello? Carry with an E at the end and illusion missing an l? And this profusion of Ys boggles the mind."

Elmsworth drummed his fingers on his thighs. "Had your ancestor known there'd be a teacher among her descendants, I'm sure she'd have taken greater care with her spelling."

I couldn't help the grin. Elmsworth grinned back, and suddenly life appeared less bleak.

"Mrs. B had serious issues with the glamor. It kept slipping," Rosie said.

Marty nodded. "Yup."

Jenna returned to her seat. "That would've been Myr's itchy hex at work. Or Mrs. B isn't a White and has glamor issues. Still, she teleported."

"Myr did too," Daisy said.

I would never hear the end of this.

Elmsworth stretched out his long legs and steepled his hands. "Whatever she might be, she's got her skylles down pat. Makes me wonder what's in Ignatius's ledger. I'm also curious if we can't perhaps persuade the guy to help us."

No need to tell me who would do the persuading. But he was right. "As much as I hate it, it's time to involve the tonker."

Mel frowned. "One thing makes absolutely zero sense, though. Two things, actually."

"The three days of grace?"

"That's one point, yep. Why give us time to regroup?"

My recent hexing stunts tumbled through my mind like rubber ducks in a mill race, and the answer wasn't hard to give. "She's not doing that in the slightest. After conjuring up tons of magical tricks, I would bet she's as drained as I am. Molly from the bank describing her as nondescript means she wasn't trying to impersonate me. Heh, most likely, she couldn't, since she's plain out of supernatural mojo. And I'd bet she won't need the full three days either to re-charge. But it'll take some time."

"That's point one covered, by why pull these stunts in the first place?" Jenna asked. "The petals, the glamors. Why frame Chris? Why bother to steal or sneak into the vault instead of going straight to extortion?"

Rosie banged her mug on a wrought-metal side table. "She's afraid of confronting Myrtle."

"Huh?" I pointed at my chest. "Moi?"

She smiled. "Yes, dear. You. You're quite formidable and still ramping up. She hates you for sure. Yet she also fears you."

"Somehow, I can't believe that. Bob Ignatius doesn't seem to have a problem with me."

"He's certainly special," Jenna said. "When are you going to talk to him?"

—

Chris's obnoxious uncle agreed to see me the next day. While I mentioned Mrs. B's ultimatum, I held back on our suspicions concerning the woman's true nature. Bob Ignatius being such a slick operator, I needed to keep an ace up my sleeve.

Agreeing on a meeting place turned into the clash of the Titans. He invited me to his villa in Bath, but did the guy seriously think I'd pay house calls to someone like him? If he did, he suffered from a serious case of wishful thinking. After some haggling, we compromised on a real estate office in Swindon, one in a nationwide chain belonging to the Ignatius business empire.

Of Chris, there was no news, other than the lawyer being what he called "cautiously optimistic."

When I went downstairs the next morning, edgy after another broken night, Alma's bombshell did nothing for my mood. "The police have taken Mrs. Bingham's flat apart. That's how we know who they suspect. Huh, you newbies are unlucky, that's what you are. First Dot and now this Bingham person. What are you doing during these parties you lot keep having?"

"Not plotting criminal acts, for sure," I said.

Cecily's expression was doubtful. "Folks around here keep wondering if you're a death cult. They say you're a Danish on the face of the village."

Urgh, that was worse than expected.

"Blemish. And thanks. We're not. We're history nuts, we share a common heritage, and believe it or not, we care for this place. A lot."

"Not sure they'll believe that," Alma said.

Whoever "they" were, I had neither the time nor the patience for anonymous haters. "Later. I've got to see someone. Do you need me to pick up or deliver anything? The sheets are already in the van. Looks like we're a bit low on guests right now."

Alma and Cecily swapped gazes. "With people dropping like flies in this place, what do you expect?"

I winced. Exaggerated the statement might be. But wrong? Oh no.

"There was a bunch of reporters, but I told them we're booked," Alma declared. "Don't need no rabble-rousers in here."

That was an interesting label for the free press.

"There's a couple of new sales reps coming in today. Word's going round about our deals, so we're fine." Cecily looked smug.

"Anything to support the working population." My fake smile slipping already, I left.

Chapter Nineteen

Arrived fashionably late for the meeting with Bob Ignatius, I then cooled my heels in a featureless waiting room until he arrived twelve minutes and thirty-two seconds after me. No doubt a clever ruse to put my hackles up, but since my hackles were up already, the maneuver was moot.

Chris's uncle was in his business persona, wearing a charcoal suit so sharp he could have cut himself on its creases. No tie, ties were out as I'd read somewhere, probably in one of those glossy rags to be found at my local hairdresser's.

Oddly enough, Ignatius's tired face didn't match his suit. Nor did he radiate arrogance, let alone menace.

He slumped in the opposing seat. "Right, your Mrs. Bingham wants my ledger."

"She isn't ours," I hissed. "And what's so important about the wretched thing?"

The ghost of his former arrogance appeared on the man's face but bowed out again. "All the insights gathered by my family back in the sixteen hundreds in one manuscript, including a very detailed description of the Whites' disappearance through the Avebury stone circle."

Talk about embarrassing. I, the witch, had to pump a descendant of the witch hunters for intel on my kin. "We assume Mrs. B plans to go through the circle herself. That's why she needs our magical plaques. If your ledger describes the process, she'd need it too."

"Why?" He clawed at his hair. "It's impossible for a Red. Beyond stupid to even try."

"She's killed twice, Mr. Ignatius."

He leaned back in his chair, a chrome and white leather affair. "I have faith in our police force."

"Oh, do you? Then why are we here?"

A derisory smile played around Ignatius's thin lips, and this time it stayed. "Insurance, Ms. Coldron, insurance. And isn't this a great opportunity to discuss all these things we never had the time for in the past?"

Unbelievable. With a murderous witch on the rampage, the guy still obsessed over business.

I'd have to show my ace earlier than expected. "Look at this." I held out my phone with Mrs. B's message.

He read. Blanched. Abject terror tore at his face. "This is impossible."

"You're referring to the possibility of Mrs. B being a White."

When Ignatius swallowed, his Adam's apple bounced up and down his scrawny throat. "It's impossible," he whispered. "They left. My ancestors chased them away and protected humankind from these horrors."

"Jury's out on who's the real horror here, Mr. Ignatius."

He shook his head. "You've no idea... With them around, no one would be safe on this planet."

"Like I feel safe with a bunch of egotistic lunatics playing fetch with atomic missiles, sure. If they're not busy doing that, they bust the planet, all in the name of profit."

He raked his fingers through hair that must've once been

dark like Chris's. "We've no time for ideology. She can't be a White. They're gone. But even if she's only"—he hooked quotation marks in the air—"a very competent Red, she must be stopped."

"Tell me something. If the Whites were Nature's X-force, how did the witch hunters ever get on top of them?"

Chris's uncle drummed his fingers on the table, a gesture so familiar it brought tears to my eyes. I blinked them away. Weakness would get me exactly nowhere.

He stared out of the window as if in search of answers. The view was nothing special, more windows showcasing people in offices, but it must have given him inspiration.

"All witches, Red or White, have what you might call an Achilles' heel. Haven't you noticed?"

"Only one?"

"One I know of. For you lot to unleash your full powers, you need to be on the ground. Better even, you're in contact with nature or natural materials. Wood, water, stone, it all works. However, none of your ilk can fly. Suspended in the air, even the Whites became pretty harmless. Not immediately, but after a while they lost their, uh...skylles. Some of them forever. I believe the hunters trapped them with nets and suspended them in cages."

The picture was too gruesome to imagine.

A lopsided grin twisted his face. "You'll work it out, should you ever try to use your magic in an airplane."

The guy was impossible. As a solution to our problem of how to keep Mrs. B in jail, his suggestion sucked, but I still asked, "How long is a while?"

"No idea."

"Brilliant. Anything else why you're at it?"

His expression serious, he leaned in. "She mustn't talk. We can't have rumors flying. Surely, you'll have a convenient spell for that?"

"I'm a magical noob. A baby joining a weightlifting contest

would stand a better chance."

The corners of his mouth kicked up. "Now, now, Ms. Coldron. I believe you underestimate yourself. Alternatively, you over-estimate the competition. Even if this person were a White—which she can't be—she'll be compromised."

"Huh?"

"I refuse to believe she's an original from the sixteen hundreds."

"What makes you so sure? And why is that a bonus?"

"First, the Earth Wardens, Red and Whites both, were mortal like you and me. Unless she's found the elixir of youth, which to the best of my knowledge doesn't exist, she must be a contemporary, possibly descended from someone left behind. To my mind, she's making too many mistakes. Then we have this three-day ultimatum, a clear sign she's having trouble with her skylles."

Since I had come to the same conclusion, I believed him. "Maybe she's getting doddery after having lived for centuries."

Ignatius threw up his hands. "Please, Ms. Coldron. This business isn't remotely funny. To demonstrate my good intentions, I'll give you this."

He fumbled in his briefcase and fished out a thin pamphlet bound in something greasy like wax cloth and encased in plastic foil. Chris's uncle then nudged it across the table with the tip of his index finger. The whole caboodle, when unpacked, contained several loose, yellowed sheets.

"Looks a bit shabby around the edges."

"It's old. Someone's personal notes, someone I suspect to have been a White."

I stared at the grimy pamphlet as if it hid a nest of green mambas.

"You might find some interesting inspiration in there. No spells, naturally." He spread his fingers and shared a grin with too many teeth in it to be genuine. "We want to catch a mutual inconvenience, not give you ideas above your station."

Thanks a bunch, you tonker.

"I won't hex for you."

He coughed. "Consider it a present. No strings attached. We have a common enemy, and I want you to succeed."

If I believed that, unicorns would fly. But I took Ignatius's gift, and when I showed it to Petty, she whirled through my living room as if we'd won the lottery.

Perhaps we had.

—

While the coven searched for a killer, trying not to alert the police to our illicit ops, Jenna, Daisy, and I spent the rest of the day poring over Ignatius's pamphlet, a bizarre collection of notes once penned by a student of the lore. The stuff in there was mind-blowing, but also patchy, chaotic, and incomplete. Nothing on blocking someone's skylles, though there seemed to be a way to "fortify" each other.

The part Chris's uncle must have had in mind when he referred to inspiration was slightly less obscure.

In a passage on "How I lyfted objects," the author revealed they didn't do any heavy lifting themselves. Oh no, perish the thought. Instead, they seemed to have trained their poor familiar. There must have once been a complete collection of sheets on the subject, but only one was left. With the text torn and parts crossed out, it took Jen, Daisy, and me ages to agree on its content. At that point we were going cross-eyed and cranky, and the event threatened to turn into a bitch-fest. So, we called it a day and staggered to our beds, but my sleep was beyond rotten, and here I was, back in my living room at the crack of dawn.

Petty dipped her fake terracotta pot onto Ignatius's document.

"Of course, sweets, you want to help." I yawned into my mug of tea. Not having found the energy to brew coffee, tea had been the easiest option.

Outside, an engine purred and tires crackled over the gravel. Headlights shone on the ceiling until they stopped and the motor cut off.

Even for the Simpkins sisters, it was too early. A sales rep having driven through the night? I pricked my ears and listened in the silence, which remained obstinately silent.

My watch insisted it was four a.m., with a summer dawn graying the furniture. Perhaps the guy delivering the newspapers? It couldn't be Mrs. B, surely, since she wouldn't advertise her arrival. Once more, I forced my sanded eyes to browse the text.

Petty gyrated into my field of vision, spinning around her own axis so fast that just looking at her made me dizzy.

"Mh, you could go invisible, sneak up behind her, and make her tumble into your pot. Nah, won't work. It's too small."

Petty hovered over the washing basket and sparked.

"Once filled with soil, it'll be too heavy, dear. Also, this idiotic document doesn't mention how long we'd have to keep her afloat. Minutes? Hours? Days? Blast it, I can't believe I'm seriously considering catching someone who might be a White. Let me read this once more."

Did I just hear the entrance door close? I pricked my ears, but the house buzzed with white noise, while the ormolu clock on the mantelpiece ticked away the seconds.

Oddly enough, I wasn't afraid. With my skylles tingling close under my skin the whole time, they would sound the alert should Mrs. B pop by. With a shrug, I focused once more on the ancient manuscript. Turning the page revealed a doodle of a brush-like object with a human body stuck on top, its mouth a big, wide O.

Petty hovered over the drawing and drooped her leaves again.

"Looks like this person's familiar was a lot bigger than you. Some sort of bush, I'd say. Hm, the idea as such is intriguing, but—"

A whisper of displaced air and a light footfall were the only signs someone had entered the room. "Blimey, you're up with the roosters," Chris said.

I swung around in my chair, and there he was. Tired and scruffy-looking. But smiling. And free.

"Chris!" I flung myself into his arms.

"Hang on." Once he had shrugged out of his heavy backpack, he returned my affections with matching enthusiasm. His cheek was stubbly. He whiffed of garlic, dust, and sweat, but who cared.

"Myr, I missed you so much."

"And I you." Words were unnecessary when I was all smiley and floaty, but heck, I wanted him to know.

As if he needed to prove anything, Chris cupped the back of my head with one hand and swooped, planting a whopper of a kiss on my lips.

Magic had nothing on him. I swayed on my stockinged feet as a wave of dizziness crashed over me. Every nerve ending in my body tingled with anticipation, and stupid things like murderous witches slipped under the surface of my consciousness without a single bubble.

Somewhat breathlessly, we parted quite a while later—only to be joined by Petty, who filled the room with her rose fragrance and showered Chris with pink sparks

"Ah, I missed you too." He fondled a nubby leaf while she stroked his hand with a blossom.

"When did you get out?"

Chris dropped onto the sofa, drawing a creak of protest. "Last evening. I tried to ring you a few times between rushing to airports, changing planes, and whatnot. Most likely, your phone was switched to silent."

I checked. Of course, he was right. Four missed calls. "Heavens, Chris. I must be the girlfriend from hell. Life here turned totally ballistic, and I... Oh, I'm so sorry."

He grinned and held up his hand. "Fret not, I figured you'd

have your hands full. Anyway, Kerluac's a total whiz. Local law enforcement, however, could've written the script for *Lawless and Disorder*. They made every mistake in the book." The laughter in his eyes died. "Sarah's giving them tons of grief, Kerluac said. Looks like she's now remote-controlling both investigations."

Guilt not only came knocking, but banging on my door with boxing gloves. Here I was planning to sneak behind my friend's back, flushing out her chief suspect without giving her a single warning.

"What's wrong?" Chris asked.

"Apart from everything? Oh, nothing much."

"I want to hear the full story, but—would have you something to eat? I caught the nearest flight from Rennes to Southampton via Amsterdam in case the coppers changed their minds. There, I hired a car and at one point grabbed myself a soggy sandwich and a quick shuteye near a petrol station somewhere on the motorway. I'll murder for a fry-up."

"Don't say things like that. If you trust my culinary skills, I'll fry you up, no problem. Want to take a shower in the meantime? Mh, maybe hear my story first?"

He sniffed his armpit and groaned. "I'll be more awake after a shower. You hit the pans and pots, and try not to burn anything. What's that, by the way? Looks antique to me." He pointed at his uncle's pamphlet and my notes littering the coffee table.

"A present from Uncle Bob."

Chris's mobile brows slanted into V-formation. "Is this the real McCoy? He's never ever parted with one of his treasures."

"I guess he had no use for this one. He was right when he thought I would. It's part of the story you have yet to hear."

Chris squeezed his eyes shut and rotated his neck. "Consider me intrigued." He blew me a kiss and hurried from the room.

The world was still bursting at the seams with trouble, but somehow, it didn't matter anymore. For Chris, I'd even rustle up culinary skills sketchier than my blasted magic.

—

Eight nicely browned sausages, a stack of decent toast, and a rather runny omelet later—I guess I shouldn't have mixed the cubed tomatoes with the eggs—Chris pushed the plate away and wiped his lips. I didn't eat more than one slice of toast, but then I wasn't hungry.

"This is beyond wild," he said.

"You're not referring to your breakfast, I guess."

"Breakfast was the first decent meal in days. No, I'm referring to your recent exploits. More coffee?" He wagged the cafetière at me.

"Nah, finish it off."

Chris filled his mug and cleared the table. "You *are* aware the woman could be anywhere in the world?"

"It's possible, but we don't think so. Assuming she's maxed out her skylles and is trying to recover, it would make more sense to hang around in the area. So far, zilch."

I rose from the chair, put the dirty plates into the dishwasher, and filled a plastic bowl to scrub the pans. They wouldn't meet Alma and Cecily's requirements, but having messed with their spotless kitchen, it was the least I could do.

"Where did you search?"

"Ask me where we didn't. The museum director even plowed through his cellar. Given the amount of junk down there, I consider it to be a genuine kindness of the heart. Oh, even Anna bothered with tossing the church. She then freaked over the rubbish she found in the Norman tub font. Or was it the bottle of whisky hidden under the altar? Vicar claims it isn't his—"

"No, Mrs. Bingham, I gather?"

"No, Mel and a few friends combed Avebury Manor. The head gardener did ditto with the grounds." I squeezed dishwashing liquid into the water.

"Holiday rentals?"

"Hah. Went through all of them, plus assorted sheds and barns. A few people are on vacation and their houses stand empty, but Alma and Cecily knew people who knew someone who had the keys, and we searched those places. Nothing. We even tried the grapevine for the neighboring villages. Still a big fat zero."

"Bugger. Do you think she might be camping? It's quite balmy at night."

"Done that too. Two coven members disguised as hikers have been scouring the countryside left, right, and center. Mel knew someone who's friendly with the shop owner, and she snuck into his storeroom. Greg swears she's not in the Whacky Bramble. Nor is she at the café. See, we've plenty of people on the ground. As have the police. But it doesn't seem to make an iota of difference."

"I take it you tossed my apartment?"

"Yup. My personal contribution to the killer hunt."

"Maybe she's using a magical trick to keep out of sight."

I threw my hands up in the air, spraying droplets and greasy suds over myself. "Then we're royally screwed. I suspect there's some place we haven't thought of yet. Search me if I know what it is."

"West Kennet Long Barrow?"

"Hah, hiker number one visited there first thing yesterday morning. No signs of illegal campers. Honestly, a damp and gloomy chambered tomb isn't my idea of an ideal hideout. No, I put my money on an empty house where she has a roof over her head but is otherwise undisturbed."

Roof. An image so faint it bordered on nonexistent flitted past the edge of my consciousness. The image returned, took shape. A slate roof peeking over trees, windows turning blind eyes on the weather.

It was hopeless. The image slipped away.

"Hello? You still in there?"

"I'm trying to remember something. Never mind. Chris, we must find Mrs. B before the ultimatum runs out, and she's recharged her blasted skylles." I used Kleenex to dry the inside of the pan, which I then returned to the draining board.

The image returned, featuring a desolate building with cracks in the plaster and greenish streaks behind the sagging gutters. But the roof peeking over the trees seemed sound.

"The empty house!" My outcry startled the sparrows chirping around on the windowsill, and they took off like feathery rockets.

"Which one?"

"The third property on Tadpole Lane, behind the farm."

"Isn't that where you found—"

"My murdered guest, Gerry, yes. The solstice killer's last victim, poor guy. Though he was in the clearing, not in the house. Anyway, it's been empty for ages. Jenna organized a key and slipped in. And listen to this. She told me the staircase had collapsed, and she couldn't get onto the top floor. That makes no sense, though. If the roof is sound, how can the stairs collapse?"

Chris shrugged. "Why not?"

Feverish thoughts raced through my head. I might be out on a limb here. Could I ignore my hunch, though? No way.

"You said yourself she might use magic. What if she glamored the steps to look decrepit when they're not?"

"You're stretching it a bit, but okay."

"It's ideal. Nothing but the farm and one other house around, apart from a few fields and woods, it's still close enough to civilization. It's where I would hide." I wadded the Kleenex.

"Please don't pull another of your Lone Ranger stunts. Start a search party. I'd like to be on it, but I need some sleep."

No wonder. The poor guy looked dead on his feet, but the mattresses in the clink were most likely crap, and spending the night in a driver's seat near the motorway wasn't anyone's idea of a good night's rest.

"You go upstairs and catch a couple of Zs as Greg would say. In the meantime, I'll see who I can rustle up. At six a.m. maybe not many people. Elmsworth might be awake, though. Or Damian and Rosie. Or both."

"Wake me when you're ready." After another too-brief kiss, he tramped upstairs.

I shot off a message and then paced the kitchen floor with Petty, waiting for a response.

Not long after, Damian's answer pinged in. We were meeting in fifteen minutes at his and Rosie's place.

"See you later, Petty." She sparked a parting salute and sailed over the white saloon-style doors, headed for the living room.

I turned the other way into the yard, leaving the back door open since Cecily and Alma were due any minute. Having rolled Auntie's orange bike from the shed, I swung myself into the saddle, and cycled off as fast as my wonky transport allowed.

Chapter Twenty

Life had moved on during my French gig—the entrance to Damian's and Rosie's house sported a shiny new door. Oxblood red, it stood out against the dingy brown brickwork of the house. Someone, probably Rosie, had been binding up the climbing roses on the porch, clearing building debris from the scraggly grass, and pruning the forsythia.

Elmsworth's Rover was parked outside the entrance, which meant he'd arrived as well. This village sheltered not a coven but a bunch of insomniacs.

I gave the door panel a perfunctory rap and entered. A soft murmur of voices, accompanied by the chinking and clattering of cups on saucers, drifted from the door leading to the living room. Perish the thought that sleuthing for murderous witches could ever be successful unless one consumed gallons of hot tea.

"Morning, all."

Damian shoved his glasses back up his nose. "Hello Myrtle, welcome. Do sit down."

I was doing just that when a snarl rose, roughly from knee

level. A sneak peek revealed Buster standing on his hind legs and giving me a full view of his chompers, all in perfect working order.

"Hello to you, too."

Buster's snarl segued into a growl, which sounded as if it should come from something five times the size of the Chihuahua.

"Buster, give it a rest," the Colonel snapped.

The day the critter obeyed, the river Kennet flowed backward.

Rosie hid a yawn behind her hand. "Ever so sorry, dear. My body hasn't heard of the sixties being the new forties and insists the scary birth date in my passport is correct."

"Don't worry, mine thinks thirty is the new seventy."

Elmsworth shuffled some papers. "I hear Chris is back? Good news, eh?"

"Yes. He helped me remember the abandoned house."

"Which brings us bang on topic. Here's an ordinance survey map. From the location, it's certainly ideal."

"Jenna found nothing apart from a collapsed staircase," Damian said with his usual deceptive mildness.

Elmsworth glared. "Could have been camouflaged...glamored, I mean."

The last things I needed were arguments at dawn. "Does Jen still have the key?"

The Colonel consulted a list longer than a roll of toilet paper. "Yes. She hasn't returned it yet. Says so here. Who'll go in?"

"Me and Petty, I guess," I said.

"Not on your own, surely?"

"Colonel, I'm not suicidal. It was dangerous enough when we were only searching for the woman. Chris said he'd join me, and I wouldn't mind another hunky man or two who can throw heavy items if required. Or someone capable of flinging spells."

"This means Jenna," Rosie said. "She does this soothing thing that makes me snooze like a cat."

"Marty will be too busy on the farm," Damian said. "I don't think I count as hunky. Nor can I throw anything far enough. Apart from words, but they won't help you."

"I fear I won't be any good at that sort of thing either," Rosie said, with quite some regret in her voice. "I'm still not fit enough. Cancer will do this to you."

The Colonel cleared his throat. "Is this the right moment to mention weapons?"

Rosie twitched in her seat. "You want to shoot Mrs. B?"

"Only to incapacitate her. I happen to own a few pistols and guns. And the required permits. I'm also an excellent shot."

News flashes about an O.K. Corral situation in the village flaring in my mind, I said, "No shooting. This operation is crazy enough as it is. If anyone raises the body count, it mustn't be us."

"So, we can't use weapons, but Mrs. B is free to hex or kill us at her leisure?" Elmsworth asked. "Doesn't sound fair to me."

"Yes. I mean, no. Jen and I will do our damnedest to dish the woman a dose of her own poison."

Elmsworth mumbled something under his breath. I had no idea what it was, but the word "civilians" popped up twice.

Rosie shook herself like a soggy dog. "Anyway, that's sorted. Back to the team. What about Daisy? She's great with Petty."

"She is. And while we're discussing familiars, there's something else you might want to know."

I filled them in on the contents of Ignatius's ledger and my, admittedly rather vague, plans to use Petty as a magical crane to lift someone who might or might not be a White off their feet and keep them hanging there until they lost their skylles.

If the situation hadn't been so dire, the variations on the bafflement theme showing in my friends' faces would have been hilarious.

"How is the bit with the lifting supposed to work?" Rosie asked.

"Not sure."

Elmsworth doodled something on a slip of paper. "Fix a plank to your familiar's pot and make her sit on it, maybe. Hmm. How long is she supposed to stay up? And will her skylles return once she's back on the ground?"

"No one seems to know, not even Ignatius."

Elmsworth pushed his paper away. "Forget it. Too many unknown factors. The other issue is we'll need a plan of the house before we can even think of trapping her."

Damian took off his glasses and blinked. Without his thick lenses, he appeared vulnerable. "If we run a recon, we might also run into Mrs. B. Catch twenty-two, I believe this is called. Don't forget, the place is rather large. Lots of corners where she can hide."

"Not if the stairs are truly broken," Elmsworth said. "She'd be limited to the ground floor."

"Gentlemen, lady, this discussion is running in circles," Rosie said. The determination in her voice reminded me of the way she handled her movers. "I say, throw a scouting team together and take a look. If there's no glamor on the stairs, and they are wrecked, we know we're wrong."

"I'm all for it," the Colonel said.

Damian put his glasses back on. "So, as for the team, we have Myrtle, Jenna, Daisy, Petty, and Chris."

"And myself," Elmsworth said. "These old bones of mine can still rattle fast enough."

"Sure," I said. "Oh, but do me a favor and stay away from your armory. Instead, bring Buster."

At hearing his name, the dog snarled. His master raised an inquisitive eyebrow. "Are you serious?"

"He's good at nipping ankles and tripping people up. Between your dog and my plant, we may be able to frighten the socks off even the scariest White."

—

Regardless of whether or not we had a key, the safest approach to the abandoned house required us to leave the farm through the back gate, follow the stone wall to the coppice, and enter the grounds via the broken garden wall. Unfortunately, this also meant confronting one of my worst nightmares—the clearing where I found Gerry's body.

The woods were okay, despite my heartbeat pounding in my ears, but as soon as the sun's rays twinkled through the branches and an expanse of grass came into view, my legs decided they'd had enough.

With only a fake acacia between me and the meadow, I shuddered to a halt.

"Sorry, folks, I can't do this," I said.

"I understand." Jenna laid her hand on my shoulder, the touch as light as a butterfly's kiss. A soothing cool flow spread from her fingers, flowing through my veins, untangling the knots deep within my innards. Buster whined and pawed my leg.

Nice try. I still couldn't make myself enter the place where my guest had died.

It wasn't as if I sensed his malevolent spirit lurking in the mottled shade thrown by the leaves, or caught his lament on the wind. Alive, Gerry had been a kind person. If there was an afterlife, surely, he wouldn't go to the dark side and come after me, the person who brought his killer to justice. Even the shreds of the white and blue police barrier, fluttering in the same breeze that tousled the grass, held no terrors for me.

The vermilion heap of bricks where the garden wall had collapsed was something else entirely, since it grew much too close to the place where Gerry's body had lain, surrounded by a cloud of flies.

The flies had been the worst. Sometimes they buzzed through my dreams.

Petty must have sensed my discomfort and nudged my hip. She arced across to the overgrown trail leading in from Tadpole Lane, the only other access to this place. From there, she sailed along the wall and hovered next to another jarring gap in the enclosure, far removed from the place where Gerry died.

"Aw," Daisy breathed into my ear. "Look at this. She's showing you how you can enter safely. Isn't that sweet?"

I forced myself to take slow, even breaths.

"Where exactly did you find him?" Chris's voice was calm and reasonable, everything I was not.

"Over there, by the butterfly bushes. I know he's not around anymore. I didn't even see him. Only bits of his clothing." The rapids thundering through my head roared and crashed. "Until I noticed the flies, I mistook him for a flower. Sorry, I know I'm weird."

"No," Elmsworth said. "You're not. You keep stumbling over corpses. It's not healthy."

"Post-traumatic stress syndrome," Daisy said, nodding sagely. "A few weeks ago, I watched this documentary where they explained it all. As the moderator pointed out, it's good the amateur sleuths featuring in crime stories aren't real people with genuine emotions, otherwise the UK wouldn't have enough loony bins to house the lot."

A laugh boiled past my lips, and that did the trick. My muscles, ready to cooperate once more.

"Thanks, I'm okay now."

A hand patted my shoulder. "We know," Jenna said.

Pursuing the route Petty had mapped out for us, we scrambled over the loose bricks and picked our way through the overgrown garden. Only the clumps of lilies poking from the weeds and the occasional sunflower soaring over a sea of tall grass spoke of a past where this place might have been a gardener's pride.

The flag-stoned terrace, cracked and darkened by fungus, also had seen much better days, as did the house itself. But the window panes, grimy and dusty though they might be, were sound, and the back door, when we finally found it, was barred shut with plywood.

"Hm, that looks solid." Chris gave voice to my thoughts. "If someone's been around, they didn't use this entrance."

"I dare to disagree," the Colonel said. He was shading his eyes with one hand and seemed to scrutinize something behind us.

"There." He pointed at the dancing grasses behind us, at a point where the brambles had taken over.

I squinted into the sun, but apart from a sunflower bobbing its seed pod, I couldn't see much.

"Ah," Chris said. "Well-spotted."

More judicious squinting revealed an erratic line where the grass had been trampled. The trail didn't start at the broken wall but emerged from a wall of bamboo plants.

"I see it," Daisy trilled and charged off.

"Dais—"

She vanished into the bamboo. With an angry rustle of her blooms, Petty zipped after my cousin.

"Ladies, gentleman," Chris said. "Watch this."

I swung around. Jen and the Colonel did likewise. Chris was tugging at the plywood until the wood, complete with door, scraped over the terrace and crunched to a stop.

"Nifty little trick," he said.

A crackle behind us had us jump, but it was Daisy, accompanied by Petty, emerging from the bamboo. "There's a rusty garden gate, propped open with a brick. Because of the plants, it's hard to spot, though once you know someone's been, you can pick out the trail."

My stomach lurched and coiled into an icy ball. Between the theory—we must catch Mrs. B before the cops did—and practice, which meant confronting Mrs. B, stretched the void between galaxies.

I wiped my clammy hands on the legs of my jeans.

"Okay, I'll go in first," Chris said.

"You're not fit yet."

"Thanks for your concern, but let me be the judge of that."

"What if she flings a hex?"

"She didn't hex me," Jenna said. "If that's Mrs. B in there."

"Hah, it won't be the estate agent. Not with the place in such bad shape," Elmsworth said. "Go in, Chris. I'll cover your back. Not you, Buster. Sit."

Chris vanished into the house.

For once, Buster wasn't guilty of assault. Trembling, with his tail stuck between his legs, he was a shadow of himself.

"I'll take his leash," Jenna offered. "He usually doesn't bite me."

He didn't seem to be in the mood for biting anyone, let alone Mrs. B. If the little guy was acting like this now, what would he do if our killer turned up?

"Eh, the dog—"

Either Elmsworth didn't hear me, or he didn't want to listen. He too entered the house. An instant later, Petty hovered after the men.

"Petty, for heaven's sake."

A spark popped from behind the plywood and winked out.

The wind blew. A fat cloud drifted in and cut out the sun. In the recesses of my brain, something buzzed, but it wasn't a cloud of imagined flies. A faint green scent announced the rise of my skylles.

"What are they doing in there?" Daisy had never been the most patient of people, but I didn't blame her. Something was wrong with this place, where time stretched from now into a future as yet unknown.

Elmsworth stuck his head back out. Time snapped back into the present.

"All systems green. You can come," he whispered.

"What about the stairs?" I whispered back.

"Hah. They won't win any design prizes, but they're in

perfect working order."

"Blast," Jenna said.

Why did I have to be right?

Once inside, we crept along a narrow corridor speckled with mold and reeking of damp pond. At its end, Chris and Petty guarded an open door, which gave a full view of the hallway, complete with staircase. Covered in the remains of a carpet, it led upstairs.

Jenna exhaled a lot of breath in a quiet rush. "This wasn't here the last time I came. I found one big heap of rotting wood, and the walls were covered in mold and moss as if the roof had collapsed. If Mrs. B weren't such a beast, I'd admire her for the glamor. I bet, though, it didn't last long."

"Long enough to fool us," Elmsworth said.

"Me, you mean," Jenna said.

"Shhh," said Daisy.

Chris tiptoed closer to the steps, and my hands clasped at the empty air, as if I could hold him back. "Don't. Booby traps," I hissed.

Chris froze.

"Petty, can you check for us?"

My familiar dipped her pot in response and hovered for the steps. The leaves and blossoms stayed upright, which seemed to mean nothing nasty and magical waited for us. Step-by-step, my plant floated up the stairs until she stopped at the landing. There she grounded her pot and fired a pink spark.

"Flight control seems to think the approach is safe," I said. "Shall we have a look or leave?"

No one laughed. No one left either.

Jenna handed the leash over to Elmsworth. "You better look after your hound. He's not happy."

For the first time since I'd known Buster, I felt sorry for the grumpy critter. "His senses are better than ours. I bet he smells or hears something scary."

"I don't need to smell or hear anything to know I'm in full

panicky chicken mode," Jenna muttered.

"As am I." Sadly, I was supposed to be the coven's top-shelf witch, so I ignored the urge to run and climbed the stairs. Not a simple thing to do with trembling legs and a throat too tight to breathe properly. Chris, who seemed to think he was on a bodyguard detail, blurred past me and reached the landing first.

"Did you have to do that?" I asked once I arrived.

He opened his mouth. Then he shut it again. "Did you hear that?"

"My hearing isn't what it used to be." Elmsworth, climbing the steps, wiped his forehead with an oversized hankie.

"I can only hear you lot talk," Daisy said.

That was when the house sighed.

Chapter Twenty-One

O nce the sigh had fluttered away, the abandoned house returned to its former quiet, broken only by the creak of a floorboard and some rather rapid breathing all around me. No one moved, not even Buster. His tail between his hind legs, the poor dog was shivering as if caught short by an arctic outburst. Yet the little guy wasn't making the slightest noise, which was seriously out of character.

The critter was more intelligent than I'd given him credit for.

Petty moved first. Silent as a ghost, she shimmied her pot and floated up the remaining set of stairs toward the banisters of the first-floor gallery.

A clammy, fetid chill permeated the air current, drifting down the staircase to tease the top of my spine.

Petty, arrived on the top floor, winked out of sight. Two muffled raps sounded from the newel post guarding the stairs.

No as in "I don't like this?" No as in "Don't move, or else?" The latter made more sense.

My cousin tugged at my sleeve. Her meaning couldn't be clearer if a question mark had blinked on her forehead.

"I have absolutely no idea what she's on about," I hissed. "But I'm not moving an inch."

A welcome fizzing and the green scent that accompanied my skylles gave me the strength to remain calm. If Mrs. B burst from a hidden cupboard and flung hexes, I'd be more than ready. I did not know what I would fling back at her, but I trusted my skylles to get creative.

The house sighed again. But it wasn't really a sigh; it was a hush of disturbed air. Someone had opened or closed a door.

Petty materialized. She bounced around with a lot of rustling and swishing and then dipped her pot onto the newel post. A single rap banged into the silence, and things clicked for me.

"Mrs. B's gone. We're safe." As fast as I dared, I sped up the remaining stairs.

"Blast it, wait." Chris dashed after me, and we reached the first floor together.

Two corridors, one left, one right, stretched into the half-gloom caused by grimy windows and no electricity.

As creature comforts went, this hideout was minus three on a scale of ten. It maxed out on the creepy scale, though. In the recent past, I spent way too much time in seriously crappy places, which was not good for morale or my stress-life balance.

"Did you spot anything?" I asked Petty.

She zipped along the corridor on the right and stopped in front of the last door at the far end. Then she whirled around and disappeared behind some sort of half wall. The next moment, she popped out again. She repeated the maneuver twice.

"Let me check." Chris headed for the wall, and I let him.

I had little choice, not with green creeping into my vision and the floor undulating as if it hosted an earthworm convention beneath the grubby floorboards. Pale pink petals swarmed the air. My skylles wanted out, and I had a hard time keeping them on the leash.

"Cool," Daisy said with envy in her voice. "You're glowing."

"I knew it," Jenna said. "The moment the recipe book mentioned blood spells, I knew."

"Knew what?"

"Not now. Just look at yourself. Amazeboggling."

I took a peek and, sure enough, a faint but noticeable glimmer danced over my hands, arms, and probably the rest of me. How odd. While the corridor wasn't exactly bright, it wasn't dark enough to require a witch light. The one time I'd conjured one up, the effects had been quite different. And painful. This time, I sensed no pain whatsoever.

It means you're doing your job. You're protecting the others.

As explanations went, this one sucked, but I couldn't think of anything better. Thinking required too much energy. And focus. Focus just wasn't part of my bag of tricks right now.

"There's a set of back stairs here," Chris hollered. "From the direction I'd say they end up somewhere close to the corridor where we entered."

"Send the primula down," the Colonel said. "I don't like this. Not one bit."

Some things are better left unsaid. With a jerk, Petty flew at Chris with such vehemence he stumbled aside, away from the back stairs. Unable to stop his momentum, he crashed onto the grimy floor.

"Oof. Yuck."

Petty landed on top of him and winked out. How outlandish. Not only did the plant, complete with pot, vanish as usual, but Chris was gone from sight, too. That was a new trick. So far, only objects *in* the pot disappeared when she did.

Then I heard it. Footsteps, slow and measured, plodded up the hidden staircase until they stopped at what sounded like the halfway mark.

Without a single word, Elmsworth handed Buster's leash to Jenna and stepped in front of her and Daisy, extricating something from his pocket that looked suspiciously like a revolver.

It says a lot about me that the sight of the weapon yanked me out of inertia when all those magical tricks, including my own, failed to do so.

"Whatever you do, don't shoot," I snapped. "I'll think of something."

"You've got enough magical mojo to blast Mrs. B all the way to Swindon," Jenna said in a casual tone. "Just envisage her there and bingo."

Swindon? For an instant I toyed with beaming Mrs. B to the cop shop, but I couldn't stomach the hoo-ha likely to ensue if a criminal materialized out of nowhere. Even if I aimed for the front door, my skylles would no doubt dump her through the ceiling or something similarly crass.

However, with the weird glimmer still coating my body rose a bizarre sense of confidence.

I'll catch her, and she'll stay put.

I sprinted along the corridor and stopped at the wall, which indeed hid a narrow and steep staircase leading downstairs, probably to the kitchen.

"Servants' stairs," Chris's voice said from somewhere close to the floor.

The skirting board gave off a gentle knock, which I took to mean Petty agreed with him.

Whoever was on the stairs hadn't moved either up or down.

How did one address a killer? And what to say? "Freeze" wasn't an option since I wasn't police.

Instead, I could try to turn her into an ice lolly. Now, there was an idea.

An even better idea would be to keep my trap shut. Mrs. B was still way ahead of us in something which wasn't a game, so it was downright stupid to give up whatever advantage I had.

A furtive scraping noise drifted from the steps, pulling away from me. Mrs. B, who seemed to change her mind more often than the average teenager, was climbing back down. Unless she'd noticed our Nancy Drew posse.

Stay where you are.

The footsteps ground to a halt. A strangled gasp sounded from the staircase.

Hey, Jenna was right. This glimmer effect worked miracles.

"Petty?" I whispered. "Can you go down? I'll try to levitate the bitch."

Chris shimmied into view, sitting on the floor, his legs extended, holding the flower in his lap.

Petty sparked and headed for the top of the stairs. There she hovered in the air, a picture of indecision.

"What's wrong?"

She shot aside, and I had my answer when a loud bang ricocheted through the stairwell, chased by a cloud of putrid green fumes, which shifted me right back to a long-forgotten chemistry lesson.

"Faugh." I too whirled around, holding my nose.

The window at the end of the corridor was closed, but Chris was already yanking at the handle and ripped it open. A cool breeze rolled in, dissipating the haze.

Maniacal laughter echoed from the ground floor, but the House of Horrors sound effects didn't sit quite right with the birdsong and sunshine coming in from outside.

Whatever Mrs. B had thrown at me didn't seem to be lethal, since I could breathe just fine. Once I'd taken two steps down the stairs, I realized the error of my ways.

There was no oxygen left in my part of the stairwell.

I shot back up and sucked in a lungful of putrid but otherwise harmless air. "I can't get down there," I said once air had returned to my windpipe.

"Why not?" Elmsworth asked. He leaned out of the window and examined the garden, the revolver still in his hand.

What part of "Don't shoot" did the man not understand?

"No air."

More maniacal laughter. This was getting old.

What did Jenna say? Something about envisaging our foe. My focus had never been clearer. I called up memories of Emma Bingham. A tall and angular woman with large hands and feet and a horsey face, she was a piece of cake to imagine.

Now, to the hexing part of the equation.

Emotions powered the skylles—check. Enough fury to heat the entire house boiled through my system.

The skylles required purpose—check. The sooner Mrs. B landed in the clink and stayed there, the better.

With that sorted, I imagined her frozen in midair.

Oh, snap. Any hexing stunt in the past involved formulating my wishes in my mind, so I'd better keep that in spec.

I want her up in the air, unable to hex.

"Petty? Be ready." To do what exactly I didn't know, but hoped she would. Then I pointed at the stairwell. *Skylles, get her.*

Brighter than a lightning flash, a beam shot from my finger. A split-second later, the backlash from releasing so much energy rammed me into the corridor wall. High-pitched ringing drilled in my ears, and my environment dissipated into a haze. For quite a while, I couldn't tell up from down. It didn't help that the floor seemed to be see-sawing under my feet.

A shot rang out. A hippo charged down the stairs. Someone screamed. It might have been Daisy. Or Jenna. Or both, but their voices were swallowed into the hubbub raging in my head. It took a while before Chris's arms around me registered. A cozy heat radiated from his body. We were both alive and safe.

"Myrtle?"

"I'm okay," I mumbled and looked at him.

He mirrored the move, concern on his face. "You sure?"

Petty bumped my hip. Absent-mindedly, I stroked a leaf.

"I feel much better. Relax, both of you." I wriggled from his embrace and glanced into the now-empty corridor. "What happened? Where are the others?"

"Let me recap. You did this Star Wars thing, and you must have hit Mrs. B, for she yowled like a scalded cat. Next thing, she racketed down the steps. Your cousin was tugging at Elmsworth's sleeve, shouting at him to do something. That's when the pistol jumped out of his hand, ending up in the air. The next second, it went off."

"Eh, what?"

"Yes, quite the sight. Jenna hopped around like an Energizer Bunny, laughing and clapping her hands. Elmsworth stood there, staring at his pistol. Fortunately, it didn't fire again but plonked to the floor. Any suggestion what happened there?"

The vaguest of notions took shape in my befuddled mind. "Your uncle's pamphlet mentioned witches fortifying each other. Perhaps that's what Daisy did, by amping up Elmsworth's magic."

"Wow."

"Was anyone hurt? Oh my gosh, Buster?"

"Alive and feeling a lot better. He charged downstairs the same moment the back door slammed shut. That was the signal for the Colonel and his troops to join the stampede, bowling through the garden in full pursuit, though I'm not sure they'll catch her. The woman's quite the sprinter."

Frustration welled up. I could have howled.

I'd flunked it.

My gaze fell to the floor where chunks of broken plaster had powdered an already dusty carpet. "Where does this come from?"

"The shot hit the wall. With a bit of luck, no one will have noticed a thing. Uh, Myrtle, there's something. Not sure if you're up to it."

Sadness had crept into his voice, and my chest tightened. "What is it?"

He inclined his head at the last door in the corridor. "The

door opened when I let the air in, and since you were busy, I had a look. I wish I didn't." He licked his lips, his gaze darting everywhere, but avoiding mine.

The next instant, I knew what we would find. And whom. And why my hex had misfired.

—

A long time ago, the room we entered must have belonged to a child. Stained and faded Mickey Mouse wallpaper covered the walls, and the bed sagging under the sash window was too small for an adult. Cobwebs festooned every corner like grimy garlands, their occupants as dead and gone as their desiccated prey. A putrid, greasy reek hit my nose the same moment I spotted the tins, empty water bottles, and fast-food wrappers next to the bed.

My heart took a plunge for the floor, dirty and covered in smudged footprints, where someone had been going around in endless circles. The footprints, however, differed in size, which meant more than one person had been here, but only one of them left.

The killer.

"Behind the door," Chris said. "It's not too bad, but you'd better not look too closely. Better call Sarah. She'll take over."

He was wrong. I needed to see this one. This one was personal.

I sucked in more or less fresh corridor air and peeked around the door panel.

Sure enough, there she was, a heap of clothes and limbs huddled into the corner, facing away from us. In places, the grayish blonde hair had been snipped close to the scalp, as if the hairdresser from hell had been at it. Dried blood pooled under the body, still giving off a rancid copper reek.

Nausea welled up, and I switched my gaze to the corpse's stockinged feet, black with dirt, a pink toe peeping from the grimy sock.

Next to one foot lay a twig. I squinted. No, not a twig but a pencil stub, its tip flat.

"Do you recognize her? I have my suspicions, but I didn't dare to go farther. The place is a crime scene." Chris's voice floated in from a faraway place.

"I recognize the hair and the clothes. She always wore olives, khakis, and taupes."

My gaze slid over the silent bundle on the floor and searched the walls until it stopped at a spot near the child's bed. Someone had scribbled on the dirty wallpaper.

"I need to get in there."

"Myrtle, you can't."

"I have to. See this?" I squinted at the writing. "She left a message. Once the cops are here, that's it."

"Unless you can levitate, you better stay where you are. Otherwise, Sarah will whack you over the head with the rule book."

Levitation was out of my league. But someone who had the trick honed to perfection floated right next to me.

I whipped out my phone. "Petty, if I put this on automatic, can you skip across and shoot some pics of the text for me, please?"

Yes, she rapped against the door frame.

The corners of Chris's lips kicked up in a wry smile. "You're the most creative witch I've ever seen."

"I take this as a compliment, even if the competition's limited. Okay, Petty? You need to be fast, since you have ten seconds before the thing goes off." I placed the phone in Petty's pot. My familiar whooshed to the wall, and an instant later my phone gave off a series of clicking noises.

Once the camera had fallen silent, the plant returned.

"Thank you, sweets."

With Chris peeking over my shoulder, I scrolled through the photos. They weren't brilliant, but once I'd enlarged them, I could make out the words.

Myrtle, I'm sorry for causing problems. She'll kill me. I can see it in her eyes. You're clever; you'll spot the pencil. I'll hold it in my hand when she comes for me. Fights are out of the question. I'm too weak now, but this I will do. Found it in the bed. I distributed the letters for her. She promised to give me what I wanted most; that's why. Stupid, stupid, stupid. Even more stupid that I told Gloria. She confronted the bitch, and it got her killed. The little beast was ranting and raving about it. There's another one dead in France, a man. She claims it was an accident. She shoved him and he fell, but that gave her ideas. Be careful. This woman hates you with a passion. I don't know her name, but I've seen her around. Don't think she's a villager. Good luck.

Emma Bingham

My hex didn't misfire at all, but it couldn't stick, not when I gunned for the wrong person.

How I wished ghosts existed, and I would get another chance to tell Mrs. B how sorry I was myself for doubting her. Now it was too late, but in her last moments, my fellow witch had risen to greatness and pointed an accusing finger at her killer, and she did so without exposing the coven.

"Oh, Mrs. B. I'll find her for you," I whispered.

"Call the police," Chris said. "You can't hide this mess. They'll take it from there."

As usual, he was right. Tears fudging up my vision, I called Sarah's number.

Chapter Twenty-Two

B itter and thin, Sarah's coffee hailed from the cop shop. At least it was hot. Together with the warming rays of the sun and a bunch of cheerful sparrows hopping around on the garden wall of the abandoned house, the brew helped drag my mood from the dungeons back up to the ground floor.

Sarah was leaning against the bark of an acacia tree, a grade-A scowl on her face. Her sidekick, swarthy, sleek Sergeant Cameron, fiddled with his tablet.

He looked up. "Okay, Sarge, shoot."

Instant panic zapped my core. Chris and I might have sent the Nancy Drew posse home with Petty and picked up the broken plaster in the upstairs corridor, but the bullet from Elmsworth's revolver was still stuck in the wall. If the scene-of-crime officers found it, we'd have a hard time explaining its presence. Since it was unrelated to matters at hand, a discovery would only muddy the waters. The fewer people dragged into this fiasco, the better.

Sarah directed her laser gaze at me. "I'd like an explanation for what you and your jailbird boyfriend were thinking when you barged in on an active investigation."

"I was cleared of the murder charge," Chris said, his voice so mild it hid a cauldron of fury.

"Plausible deniability, Mr. Lentulus," Sarah said. "There was another person around, so it's unclear who killed the private dick. Fine. Strike the jailbird, Cameron."

As Mrs. B's last message claimed, the PI's death had been an accident, and no doubt Sarah had read it. Still, she let Cameron take our statements and kept us waiting for over an hour.

"Didn't write it down, Sarge. Figured you might be venting."

"I was. Myrtle?"

I gulped and sat upright. A slump wasn't a good starting position to launch my defense. "Sarah, you're being unfair. As I explained to Constable Cameron, Mrs. Bingham was the second person in our group suspected of murder. After Dot, we couldn't just sit around and knit."

"You don't knit."

"I wish I could. Anyway, you don't have the resources to search every single empty house in this village. And even if you suspect something, you'll need a warrant. We don't. Be honest. Would you have found her as fast?"

Sarah glared. "That's beside the point."

"No, it isn't," Chris said. "You have a killer on the loose. Again. You need help."

The constable waved his tablet. "Sarge, he's right. Well, he isn't. But—"

"Cameron, I prefer to run my investigations myself, thank you very much. We can't have amateurs barging in on a crime scene."

Cameron slipped me an apologetic grin. He might be police, but he was good people. So was Sarah when she wasn't hopping mad.

I jumped up. Time to break out the big guns. "We got nowhere near the body. Your precious crime scene is pristine.

Be grateful we found her and not some kids playing house."

Like Jenna's twins. My stomach knotted into a spiky ball.

Sarah groaned and massaged her scalp. "Aw, heck. Tell me something I don't know. By the way, your Mrs. Bingham left a cryptic last message." Her gaze drilled into mine until I was sure she could read my mind. Which was daft, when even I, a witch, couldn't do such tricks.

"Ah, and you need my help to make sense of it."

"Like this, it'll go quicker. It appears she posted letters for the perp. I suppose she's referring to the ones you mentioned. Does that surprise you?"

"In hindsight, no. It had to be someone local helping the killer. I mean, teleporting hasn't been invented, correct?"

"Of course not."

As long as she believed that, we were fine.

"Mrs. Bingham claims she was promised something for services rendered, something that mattered a lot to her. Would you know what it is?"

The truth and nothing but the truth was out of the question, but the closer I could stick to it, the better.

"Told you it's that history thing. She obsessed over her heritage. The killer must have promised information I couldn't give her."

Chris let rip an odd noise, a crossbreed between a huff and a sigh. It cost him a frown from Sarah, but thankfully she didn't engage.

Cameron, who had been tapping away at light-speed, looked up. "Death by history? That's a new one."

Sarah rolled her eyes. "The victim wasn't killed because she got her kicks out of the past. She'd seen the killer's face. This person is a total fiend, I tell you. Hacked off the poor woman's hair in a hissy fit."

Not a fiend, a witch, someone who needed the hair for the impersonation hex. Another intel I could never share with Sarah, so I made sure to look shocked. "Heavens."

"Here's another one for you. Mrs. Bingham claims she's seen her killer around, but she doesn't seem to be a villager. Any suggestions?"

I took a mental note of the pronoun. "So, we're talking woman?"

"For the time being, let's proceed with this assumption. We'll also assume that she and the person responsible for the French debacle are the same. You established she has an American accent and, at one point, wore the victim's bucket hat."

Great to hear I was useful after all.

Sarah stared at the trees soughing in the wind. Then she relaxed her stance. "You better be careful. Mrs. Bingham left a warning. This person hates you. I have this hunch she might even have orchestrated his arrest,"—she raised her chin at Chris—"which means she has it in for you both."

She wasn't telling me anything new, but it was kind of her to try. "You just made my day."

"Sorry."

"It's okay. As long as I know. How...how did Emma die?"

Sarah massaged her temples. "Shot."

A memory of the coppery reek filled my nose, and I fought a cresting wave of nausea. "Thank you," I whispered, once I had forced down the bile. "I guess that's it. I better return home and get to grips with this."

"You do that. Go home and stay there. Think of anything you might have seen or heard, and should you get hit by inspiration, call me. Do not, I repeat, not go after a killer. You did that once, and it nearly went wrong."

"I didn't know I was going after the killer."

"This time you do, so kindly leave this business to us professionals."

Chris rose and held out his hand to help me up. "We'll do our best. Should the killer chase after us, we might not have the time to call."

That left even Sarah speechless.

—

The short stretch from the murder house back to Tadpole Lane passed in a blur, without either Chris or me saying a word. My suspect turned out to be another victim. With all the clues leading nowhere, how could I expect to ferret out the killer?

"Careful." Chris bent a low-hanging branch out of my way and stopped. "Did you mean it when you said you want to go home?"

"What else can I do? After this morning's disaster, I need to switch off before I can face the others again."

"That's exactly my point. The last thing I need is your cousin in freak mode, and I'll double my lawyer's fee if this upset person isn't her."

The high-pitched complaint drifting in from somewhere in Marty and Jen's apple orchard sounded a lot like my cousin. "I guess we better take Daisy with us. And Petty."

I took a step toward the entrance but stopped again. A true coven leader would charge to the rescue and pour oil on troubled waters. Not having much oil left to pour, I fidgeted where I stood, hating the inner voice that jabbered on about things like "duty" and "responsibility."

Something, anything, which gave me a reason to run the other way. "I need some distance. And a decent coffee. Sarah's brew was the pits."

As an excuse to turn my back on yet another disaster, the comment sucked something chronic. Jenna's coffee was delicious, and she'd throw in a homemade biscuit, if not a cake.

"True, and I can see you're not keen on drinking it here. Home?"

"It's where Alma and Cecily are. They'll be in the middle of turning the house inside out."

"How about the pub?"

"Great idea. Let's."

The comment sounded a lot more cheerful than I felt. I'd failed Emma and Gloria both, and now they were dead.

We turned and walked off in silence, which gave me more time to mull. The long and short of it was that the killer, once more without a face, had tricked us all. She was free to ramp up her magical mojo and continue where she left off.

"Do you want to talk rather than pull faces the whole time?" Chris asked.

"I would if I knew what to say. Who is this person? What's she got against me? And where does she suddenly spring from? It's like one of these puzzles where you have nothing but blue sky and none of the pieces fit."

"In that case, you need to start with the clouds," Chris said.

"Har, har."

By that time, we were tramping over the footbridge leading over the river Kennet, gurgling and splashing below the planks. Where the sun's rays hit the water, they splintered into gazillions of sparks. Cool and moist, the scent of the river reminded me of my skylles, but they'd gone into hiding.

I couldn't blame them.

Note to self: Call Jenna about the odd glow on my arms. She knows something and isn't telling.

Chris sighed. "Talk to me."

"I know I'm in a bratty mood this morning."

"As if I was blaming you. There must be something here, though. You're just not seeing it. For one thing, I find it interesting that Mrs. Bingham claimed she'd seen the woman around, but didn't recognize her as a villager. To me, it shows she's spotted her more than once, so she can't be any of the day trippers. She must be some sort of regular."

"A Pagan? There's tons of them; how am I supposed—"

"Wait. Pagans are one possibility. There must be other people who show up more often."

Like the trout now breaking the surface of the river in a silvery spray, a thought flashed in my mind, only to splash back down again.

"Hikers?" No, that wasn't it.

"The weekend visitors?"

That wasn't it either. "It's like winning the lottery. One in an eighty million something chance to make it."

Now Chris pulled a face. "That's defeatist."

"What am I supposed to do? I made too many mistakes."

"Since when are you responsible for everything that goes pear-shaped in this place? You've been doing overtime recently. I mean, Darth Vader would be running for safety if he'd seen you this morning."

"More likely he'd have a good giggle. Gurgle, sorry."

"Woman, you *are* cranky."

"Yes, I am. So what?"

"Once again with feeling. There must be something you know and can't remember because you're too busy running this mea culpa routine."

"When I'm out of socks and should rather be running a load of washing, you mean?"

"Myrtle—"

His voice was sucked into an insistent whisper that started somewhere in the dusty recesses of my brain, in the part dating back to the dinosaurs or whenever, the part in charge of ensuring basic survival.

I ground to a halt. Stared at a precise line of fresh mole heaps.

"Myrtle, are you still in there? Recently, I'm not sure."

"The stocking."

"Needs a wash. Like mine. Laundry service in the clink is sub-optimal. Though if you ask me, I don't think household chores are our biggest problem now."

"I'm referring to the Christmas stocking. Brand-new, so it doesn't need a wash."

Chris kicked a stone, which flew straight at the nearest molehill. The top of the soil shifted, and a black-coated furry creature poked its nose out.

"You disturbed the poor bugger."

"I feel rather disturbed myself."

The mole scrambled back to its underground lair, but the whisper stayed with me and morphed into a thought. "I found it. The piece of the puzzle. When the witch searched my house for the key, she hexed the lock of the living room into a stocking. Tiddles gave her the mother of all frights, so she skedaddled in a hurry, leaving the stocking behind. Didn't I mention this?"

Chris aimed for another stone but at the last moment placed his foot back on the ground. "You mentioned her searching for the key and the bit with Tiddles. I must admit, I was wondering how she got into the living room, but thought someone might have left it open."

"Not me."

"No, but it might have been your cousin."

"She was with me in France, remember?"

"Ah, true. Sorry, I had other things on my mind."

When it came to excuses for temporary amnesia, a murder charge no doubt counted double.

"Where is this stocking now?"

I fished my phone from the pocket of my jeans. "The Simpkins sisters kept it."

The call rang through and, a nanosecond later, Alma's voice droned into my ear. "Witch's Retreat B&B, good morning."

"Hi Alma, it's Myrtle. I've a rather odd question for you."

Static buzzed, filling my inner ear with white noise while Alma remained silent. Were my questions always that odd? "The stocking you found stuck in the door the other day."

"Yes?"

"Do you still have it?"

"Yes."

That was typical. When I wanted the woman to talk, she switched into word-saving mode. "Here?"

"Yes. I can get it out for you."

"Please. Do check if it has a label."

"Label? It was good quality, not one of them acrylic thingamajigs."

I caught myself pacing around in circles, skirting the nearest molehill. Once we were done here, the poor rodent would suffer from a migraine.

"I mean a manufacturer's label. Or a logo."

"I'll check."

The phone clattered onto something I hoped would be the draining board, not the sink.

Chris's brows did that V-thing Nosferatu would have been proud of. "Why would a hexed stocking have a label?"

"Because you need to visualize the items you wish to hex. The clearer, the better. It's much easier to envisage something you've seen, something real."

"Ah, crafty."

Thinking about it, yes, it was.

With another clatter, Alma was back. "Heh, looks like you're right. There's a little square sewn onto the top. Says Jamey's Knitwear."

I'd heard of that company. Unfortunately, I couldn't remember where and when.

"That's nasty," Alma said. "I wonder how they did it."

"Did what?"

"Steal this thing from her. Actually, now this is weird."

My mind whirled. "Alma? What—"

"Looks kind of odd inside. The stocking, I mean. It's all gray, as if the pattern doesn't come through. Maybe it's lined. No, it's not. Strange."

Not weird at all, simply another proof the stocking was magical debris. The witch had visualized the stockings on the outside but didn't bother with the rest.

"Alma? Who did this belong to?"

"Huh? Mrs. Shuttlecock, of course. Told you, she's with Jamey's these days. Gotta go; the phone at reception is ringing.

I'll keep the stocking for you. Bye."

Neither a tourist nor a Pagan—Iris Shuttlecock, my friendly sales rep, was the witch.

Chapter Twenty-Three

"Alcohol on an empty stomach is a bad idea." Greg, the landlord of the Whacky Bramble pub, deposited his serving tray. "You're white as a sheet, Myr. You need food. And your man needs something to quash the taste of that doggone French slammer. Here's some roast beef and Cheddar sandwiches. On the house. Your brandies will be right up."

He tromped back to his bar, busy with the lunchtime crowd. The place echoed with voices and laughter, but most people headed back outside where the wooden benches and tables were going like hot cakes. We'd chosen our favorite table in the restaurant part of the Whacky Bramble, the two benches inset into a bay window and smothered in colorful taffeta cushions. Despite blurring the view, the diamond-shaped window panes let in plenty of natural light, which muted the effect of the tea lights flickering in deep purple bowls.

Not that I cared one iota for the ambiance, but "our" table was far enough away from the rowdy crowd to give us the privacy we needed.

"Once you've eaten something, you'll have to bite the bullet," Chris said. "No way can you keep your brainwave from Sarah. And yes, I heard you on the proof the first time. I also understand this is kind of delicate because of the coven. If we put the right spin on the story, though—"

"It doesn't matter what yarn we spin. As soon as anyone—especially the cops—puts on the squeeze, the woman will tell the world about us. I know, I know, the risk as such isn't new. When I thought Emma might be our killer, I hoped—"

"That she wouldn't talk because of being a coven member? Fat chance."

"It's moot anyway. Emma's dead."

Chris, who had taken a hearty bite from his sandwich, rolled his eyes and swallowed. "Isn't this what your aunt wanted? To go public and help society?"

Unbelievable. He couldn't be that dense. "Dot and auntie's fantasies and reality are two different pairs of heels. Even if someone believed us, it wouldn't take long for some mega-sinister people to have a wet-dream about super weapons. With the world being the way it is, you're not claiming we'll be allowed to roam free and play Robin Hood?"

He sighed. "Probably not. Though I'm still not 100 percent convinced one stocking makes a killer."

Chris had his back to the bar, so he couldn't see Greg, his bald pate shining under the spotlights, weaving his way through the crowd.

"Psst, incoming drinks."

Arrived at our table, Greg handed us two glasses of his best cognac, joined by two cups of coffee.

"Thanks, mate," Chris said. "I feel like a human being again."

Greg's grin formed a white crescent on his dark face. With a hand the size of a baseball glove, he thumped Chris on the shoulder. "Next time, run like the clappers when the sheriffs come after you. I got arrested once in my misspent youth, a drunk and disorderly. The night in the hoosegow was an eye-opener. Never again. Enjoy, guys."

Built like a linebacker, the Whacky Bramble's landlord had no problems bulldozing his way through the patrons besieging the bar.

I wrapped my hand around the coffee, inhaled the aroma of fresh Costa Rican roast, and the world was a marginally better place. Once liquid manna warmed my stomach, the outlook appeared even less bleak. "Ahh, wonderful."

"About your Mrs. Shuttlecock..."

The cup back in its saucer, I said, "It's a perfect fit. I never had her on my radar, but now I do everything slots into place." I lifted one finger. "First, the stocking. It's the biggest clue, obviously. Second, she's from New York, which fits right in with the American accent Ella—she's the Frenchwoman who sold the potpourri—mentioned." I extended another finger. "Third, and this one's a clincher, she and Emma knew each other, which must have made it easy for her to chat the poor woman up."

Chris lifted an eyebrow. "Knew each other how?"

"The Colonel told me she worked as a sales rep at one point. Somewhere up north he said. That's Mrs. Shuttlecock's patch. Maybe they even worked for the same employer."

"Hence this fixation of hers to call you 'luv'."

"Yup." I stretched out another finger. "And here's more, though it's a bit of a stretch. The last time she stayed with me, she arrived at very short notice. She must've decided her presence was required here, the reason she teleported from Carnac."

"Whoa, there's enough elastic in your theory to kit out a trampoline factory."

"Sure, but it makes sense. She might not be a local, but she's been in and out of the Witch's Retreat as if greased, so she knows her way around. Crap, she must've been sniffing me out, and I never once noticed."

"Don't start this mea culpa rubbish again. She tricked

everyone, starting with the bedbugs and not stopping with the pros."

"No bugs in my place."

"I hope not. How were you supposed to know the enemy lurked under your roof, huh?"

A good question. She radiated only ever friendliness, not hate. A masking charm? Something that slipped when she was high on emotion, like she must have been after Poussin's death?

"Apropos roof." Chris sliced into my musings. "Where could she be now? Even if she stayed at that horrid house herself, she's now out of a shelter. The place is cop central."

"She's not with me. Hm. I wonder..." I fished for my phone, but Chris clasped my wrist.

"Before you do anything else, eat something. If you carry on like this, you'll go translucent."

The last time that happened had been during the curse-induced nightmares. Not an experience I cared to repeat, so I bit into my sandwich and discovered to my big surprise I was actually hungry.

Cognac and coffee drunk, sandwich eaten, my brain was whirring along once more. I whipped my smartphone out.

"Who you gonna call?"

"Linda. If I were Mrs. Shuttlecock, I would want to be as close to the action as possible while ramping back up on magical mojo. She can't know we've made her. She'll think she has us tied up in knots. For me, there's one logical place where I'd lie low, and that's the Crystal Dawn."

Two minutes later, we learned Mrs. Shuttlecock was indeed at Linda's place, having bluffed her way in with a sob story about not wanting to inconvenience me all the time with her short notices. At the moment, she wasn't in her room, and her car had gone. Once I shared my suspicions with Linda, she let rip a volley of supremely unsuitable comments and told us to crank our behinds into gear and help search the woman's room. Which meant that visiting the pub had been a super stupid idea.

Between Chris's Beemer stranded in France, Linda's transport refusing to start, and the village not offering anything remotely looking like a taxi service, we would have to hoof it.

—

Unbefitting its airy-fairy name, the Crystal Dawn was a dark brick building crowned by gray slate tiles. Chimes jingling in the wind and the glittery beads strung over the bushes were trying hard to make up for the architectural failings, but they had a difficult time of it.

The same moment Chris and trudged up to the porch, panting—well, I was—the front door of the B&B slammed open and Linda popped out, shielding her eyes against the sunlight.

"What took you so long?"

My fellow witch wasn't the easiest of people. "Well, the Whacky Bramble isn't exactly around the corner, you know? We came as fast as we could."

Linda huffed and waved us in. "I was worried she might turn up, and then what would we do? But she hasn't. She's in my smallest room, so it shouldn't take long to search. Follow me." She climbed a flight of narrow stairs.

"I doubt we'll find much," Chris said from behind.

"Party pooper." Linda jingled a bunch of keys from the pockets of her black combat pants and rammed one into the lock.

"More like a realist," Chris responded. "In her position, I'd either take the incriminating evidence with me or I'd put a spell on it, so it can't be seen."

The hand on the doorknob, Linda hesitated. "Rats. I didn't think of that."

Neither had I. "We're here. We might as well look."

"Ladies, please. What I was trying to say—if you find nothing at first sight, search for magic."

"I'd need Petty for that." A small pang tore through my chest. Once I'd sent her off with Daisy, I hadn't thought of my familiar once, never thanked her for the great job she did in the derelict house.

Oh, well, Petty was a kind soul. She'd understand why I neglected her over coffee, booze, and the best sandwich ever.

"We'll just have to wing it." Linda threw open the door to a tiny but charming room in muted green tones with lace curtains and antique furniture. The place seemed pristine; only the pair of fluffy slippers under the bed, the black suitcase in the corner, and the latest Nora Roberts on the nightstand betrayed Mrs. Shuttlecock's presence.

Chris strode to the luggage and tested the lock, which popped open on the first try. "Ugh, I draw the line at pawing through a woman's underwear."

"Don't be daft," Linda said. "She's a killer. I'll check the bathroom."

"Hairbrush," I said.

She raised an eyebrow. "Are you serious about this hair and nail bollocks?"

"It seems to work. I have the stocking, of course, but the Simpkinses seem to be rather fond of it and might react miffed if I burn it or something, so..."

"Mph." Linda barged into the bathroom at a speed which had me worried, but apart from muffled clattering and cussing, there were no alarming noises.

I dropped to my knees and peeked under the bed. The floor wasn't up to Alma and Cecily's standards, but it was clean enough. No dust bunnies or forgotten tissues. Unfortunately, I spotted nothing of importance either, so I sprang back up from my crouch.

"Myrtle?" Chris wriggled a stocking, which made the one we had in custody back at the Retreat look like a poor relation. No wonder when the article in his hand was the real McCoy, not a mind copy. Knitted of fluffy wool in purples, turquoise, and gold, it should have looked gaudy, but it didn't. As Christmas stockings went, this was the Rolls Royce version. From the seam poked a small white label with red writing on it, proclaiming this to be a product of Jamey's Knitwear.

"Check the inside."

"Huh?" Chris peeked in. "Not quite as nice, but that's normal, isn't it?"

"Let me have a look." I took the sock from him and checked it . Sure enough, there was no gray whatsoever. This was a proper stocking, produced in a factory and not in someone's bizarre imagination. "Our stocking looks different on the inside because she imagined only its exterior."

Chris shook his head. "Amazeboggling, as Jenna would say. There are more, by the way. She has a bag full of things."

"Stick with this one. In case we need evidence."

"Since this isn't a murder weapon, it proves she was inside your house, nothing else."

He was right, of course. "Take it anyway. Is there more?"

"Nothing relevant. And I checked the underwear."

Linda reappeared from the bathroom. "No brush. No nail clippers either. Just some toiletries. The woman seems to be a neat freak."

"She digs glamors, and she isn't stupid. Worth a try." I turned full circle, taking it all in. The bed, the nightstand, the suitcase, the wardrobe...

The wardrobe.

I scooted across and flung open its door.

Apart from a few hangers, it was empty.

"Blast it. There's got to be something."

"Unless she ferries her junk around in her car," Chris suggested.

Linda sat in the room's only armchair, squeezed into the far corner. "She must've expected to be back, otherwise she wouldn't have dumped her belongings. Mind you, leaving those stockings is already pretty dumb. Makes me wonder if the woman has all her socks in the drawer, if you know what I mean." To underline her point, she circled her index finger next to a temple.

"As you said yourself, she's a killer. Her brain's definitely wired the wrong way." Chris had folded the stocking and slipped it into one pocket of his leather jacket.

"She always struck me as being quite sane, but let's not quibble about this now. Anyone got any suggestions on how to search a room for magic?"

Bland looks from two pairs of eyes told me not to expect much from my companions. Sure, Petty would spot in a jiffy if something was amiss, but while she wasn't exactly miles away, any minute we waited was a minute too long.

I sank onto the bed.

"Hey, I made it this morning."

"If you are worried about your linen, get out of that chair," I snapped. "I need to focus, and I can't do it standing up."

"Jeez, you're a tad touchy today, eh?"

"Yes."

At least Linda was savvy enough to shut her mouth. Right, this couldn't be more difficult than the other magical tricks I'd pulled. My emotions were all cranked up. I had a purpose. I wanted to check if there was a spell on the room or on the objects. Or something else of a magical nature.

Unless they'd done a total bunk after this morning, my skylles shouldn't be difficult to call back. In that case, we'd better call Petty. But at least I should try.

Right, what else? Ah, first directive. On that front, we were probably okay since I was trying to protect not only the coven but the general populace from a killer. Unlikely it would cause magical fallout.

What was I waiting for?

I want to know if there's a spell, curse, glamor, or whatever on an object in this room or hanging around in the room itself. If there is, I'd like to have it lifted.

A heartbeat passed. Then another one, followed by a tremor running through the floor and a hissing noise that reminded me of a shaken bottle of fizzy water. Green washed over my vision, and a burst of pale pink rose petals rushed in from nowhere.

"Wow," Linda said, her eyes wide.

"Isn't she great?" Chris said. "Doesn't even flinch and throws magic around like Baby Yoda."

He must have watched Star Wars re-runs in jail; there was no other explanation for his latest choice in similes.

It wasn't easy to see much through the rose shower, but from my vantage point on the bed, nothing extraordinary seemed to have happened. The fizzing died down. The petals landed on the floor.

Don't leave them lying around.

I wouldn't. But first—

"The book." Chris jumped up and pointed at the nightstand.

I twisted around in my seat and, sure enough, the Nora Roberts was gone. In its stead, a scruffy manual bound in leather lay on the glass top, the panels hanging together by a few threads.

Diarie, it read on the front panel. I stood, opened the tome, and gingerly flipped through the stiff pages, which reminded me a lot of the Coldron and Wytchett grimoires. The entries were dated, so duh, yes, this was most likely someone's diary.

When I spotted the name Lily Coldron, I drew a sharp breath. "The author of this epic knew our Lily. My ancestor from the seventeenth century, I mean."

"Then we'd better take it," Chris said. "Though I'm sure she'll notice when she returns."

Linda snorted. "She's no longer welcome at my place. I'll pack her belongings, and she can pick them up from the storage should she feel like it. We need the diary, though. I'll eat my daughter's algebra homework if this thing doesn't contain the answer to our questions. Why don't we pop over to your place and read it there? I'll feel a lot better with her crap out of my house, but I'm dying to know what's in it."

"Chicken, bock, bock," Chris said.

"Bah."

I kneeled and picked up my petals. "What are we waiting for? Petty must see this. Plus, this place is giving me the creeps."

Linda thrust out her chin. "Oi, I had it refurbished in the winter."

Still kneeling, I counted to five. Then I said, "I wasn't bitching about your interior design."

"Ah, sorry."

Some people. "Would you have a towel to protect the book? And a bag?"

Linda was already halfway out of the door. "I have both. Even better, I have the latest Nora Roberts. We'll put it on her nightstand, and should she show up before I can sort her out, our mutual friend will have some fun working out why the hex doesn't reverse, heh."

Linda might not be the world's biggest charmer, but she made a formidable enemy.

Ping. That had been the phone in my pocket. I fished it out and found a message from Daisy.

I read it once. Then I read it again.

Its content hadn't changed. The world span on its axis.

Once again, my panicked gaze skimmed words that sliced straight into my chest. I couldn't breathe, couldn't speak.

Chris flung up his hands. "Never a dull moment. What's wrong now, for I can see something is."

My chest hard as granite, I shook my head and handed him the phone. He read, and his eyes widened.

"What?" Linda asked.

"Daisy wants to know where Petty is," Chris said. "She hasn't seen her since the house."

Chapter Twenty-Four

Petty was gone. Last seen when Daisy, the Colonel, and Jenna returned to the farm, she must have slipped away with no one noticing. Perhaps she spotted something she wanted to ferret out. She was such an inquisitive little thing, worse than a cat.

She wasn't in my living room for sure. Linda's car having started on the fifth attempt, she drove us to save time. Nor was my miracle flower at the farm or in the grounds of the abandoned house, at least not in those parts the Wytchetts and Daisy could search without stumbling over Sarah's minions.

This was my worst nightmare all over again. I'd lost my Petty.

"I thought she wanted to be with you," Daisy wailed over the phone. "One moment she was there, and the next she wasn't. She was, like, mega relaxed? I thought nothing of it."

I could have said a lot of things, like "Why didn't you check?" or "Why did it take you so long to tell me?" But the words wouldn't come, not with the scream quivering under my sternum.

So what? Words were meaningless when Petty was gone.

Chris was holding me in his arms, stroking my hair, but he might have been miles away. "Shhh, it's okay. We'll find her."

We wouldn't. I knew we couldn't.

Mrs. Shuttlecock had her.

"Come on, chin up," Linda said. "We don't know we have a problem until we know it." Seated in Auntie's armchair, she leafed through the pilfered diary and kept shaking her head. "This is unbelievable."

"It might well be," Chris said over my head. "But until Petty is safe, it doesn't mean a thing."

How? How could we ever get her back?

As if to mock me, my phone pinged. The scream stuck in my chest slipped, enough to let the words squeeze past my aching larynx. I slipped from Chris's arms, thumbed the message open, and blinked at the blurred text. "That's her." My voice sounded alien, robotic.

"I have your flower, Coldron. If you don't want me to dump weed killer on the stupid thing, come to the henge at midnight. Alone without your pals. I don't give a toss about your parody of a coven. You and I have unfinished business. Iris Greene."

Apart from a fly bouncing and buzzing against the windowpane, the room was dead-quiet.

The obvious question rushed up my throat, but Chris was faster. "Who the heck is Iris Greene?"

I could have indulged in a crying fit. Or torn down the curtains and stomped on them. Or something. None of that would have helped. "We've got it wrong. Again."

Silence fell once more. The fly buzzed and buzzed until I could take it no more and slammed open the window. Naturally, the stupid insect zipped away from me and fresh air, disappearing behind the curtain instead.

"Actually..." Linda said.

"Actually, what?" I snapped. I wasn't being fair here, far

from it, but this Greene person had Petty, and we didn't know diddly-squat about her identity. Apart from one thing. "Mrs. Shuttlecock is also called Iris. But that could be a coincidence."

"That's my first point," Linda said, her voice laced with untypical mildness. "Also, this diary here was written by a Peter Greene."

Hope unfurled a fragile petal. "You mean..."

"Yeppers. Must be an ancestor of hers. Why else would she bring it? Like with your Lily and the recipe book, eh?"

"If not that, maybe the guy means something to her, and she took on his name," Chris said softly. "

"That's another possibility," Linda said. "Whatever. I'd swear on Buddy's grave—he was my Labrador in case you wonder—that Shuttlecock and Greene are the same person, and she's just trying to cause more confusion and headache. She seems to like that sort of thing."

Hope unfurled another trembly leaf. And another one.

Linda and Chris might have a point—no, no—they *must* be right, because of the diary. The diary meant a bomb to that bitch. Otherwise, why drag it around?

Hope swelled on a fragrant rush, but something hotter and sharper rose in its wake.

If the old tome was that important, our case wasn't hopeless; we possessed a bargaining chip. We could and we would fight back.

"She's going to regret this." Having squeezed the words past my rigid jaw muscles, I hacked words into my phone's screen, every single one powered by the fury raging in my chest.

"We have your diary, Greene. If you don't want me to burn it, leave my familiar alone. I don't care two hoots about your stupid vendetta. Stop this shit now or live to regret it. Myrtle."

There. I'd sent it.

Chris gave my hand a gentle squeeze. "I knew you'd snap out of the blue funk and get angry. You always do that."

"Don't patronize me."

"I'm not. I'm stating a fact."

Linda, her cheeks flushed, looked up from her reading matter and grinned. "Give it a rest, Myrtle. Your man is right. You do it every time. Mope first, get angry afterward, and always sort things out. I prefer to skip the hurt and get angry straightaway. Like that, you don't waste any time. Right, would you like a summary of the bits I read?"

"Is it even legible?" Chris asked. "These old manuscripts can be a real bummer."

"For a text written such a long time ago, it's okay. And since I'm an antiquarian, it's easier than you might imagine."

Chris raised an eyebrow. "I wasn't aware."

"Nor was I," I said.

The heat in my chest spread. Linda never dropped a word about her training. Every single time Daisy and I struggled with our grimoire—historic manuscript or not, Lily's handwriting was beyond atrocious—Linda could have helped us.

You didn't ask for help.

That, of course, was true.

"You know it now. Anyway, the gist of this is Peter Greene was a White who taught the youngsters of the coven, Lily Coldron among them. She must have been his favorite student. When the Whites fled through the stone circle and left the Reds behind, he stayed to protect them."

"Oh, wow," Chris said. "So, they weren't all selfish tonkers."

"Nope. He certainly had a cool sense of humor. I like him." Linda's eyes shone. Then she pulled a face. "He doesn't deserve this wacko woman as his descendant. That's assuming they truly are related. I haven't got to it yet, but he says in the beginning he's going to recount what the Whites did to slip through the circle. And where they went."

Her comment pierced straight through the seething bubble of worry and anger surrounding me. "Why the heck does Madam Shuttle...Greene need Ignatius's text?"

"No idea," Linda said. "Perhaps the information is lost.

There's quite a bit missing. If it's not that, she might simply enjoy stirring up trouble for everyone because the poor baby doesn't enjoy being in this world, boo-hoo?"

"Join the club," I said.

My phone pinged.

"Don't you dare, or your familiar is compost. You have the luck of the devil. You shouldn't ever have found it. Tonight, at midnight at the henge. Be there with my diary and the rest of my stuff. Or else."

My finger hovered over the phone to hack out a response, but I withdrew it. If this weren't serious, it would be laughable, so kindergarten.

Chris, looking over my shoulder, said, "She seems to think she's being original. To me, she sounds like the archetype trope villain. Minus the mustache."

"Hairs on her teeth?" Linda suggested.

I balled my fists until my knuckles cracked. "I don't care what she sounds or looks like. As long as she leaves Petty alone."

"Oh, she will," Linda said. "Not only because of the diary, but because your primula's a familiar. I don't think she's got one. She'll try to trick you and keep her."

The kick of a mule into my gut couldn't have hurt more. "Can she?"

"No idea. Ask Jenna; she'll know. I doubt it, though. Doesn't the recipe book have something to say on the matter?"

"Nothing much in the visible bits. As for the rest, I never got around to check. Too much going on."

Linda shot me a wry smile. "You can say that. Let me help you. I'll read this through to the end, and then I'll give you chapter and verse."

The weight on my chest lightened somewhat. Not much, though. "Thanks a bunch. Take a copy while you're at it. We might have to trade it, but I want to know what's in there. It might even tell us why she despises the Coldrons."

"I bet dearest Peter had something going with his star pupil, and that's why he stayed."

Linda was impossible. On any other day, I might even have laughed.

"Now that's sorted, what do we do next?" Chris asked.

"Now I organize a rendezvous at midnight with my archenemy."

"Got a plan?" Linda asked.

"Nope. My hexing's best when I just let rip, so that's exactly what I'm going to do."

"You'll need backup," Chris said.

"Colonel Elmsworth with the revolver at the henge? I think not."

Revolver. My mind flashed a loopy vision of a revolver hovering in midair, followed by a collection of yellowing sheets of paper. Now, there was an idea.

Ignatius's moth-eaten manuscript had been useful, after all. I wouldn't be coming alone. Instead, I'd bring along my secret weapon.

Chapter Twenty-Five

Another bad weather front was moving in from the west, spreading a luminous haze over the waning moon. Here and there, stars twinkled through the mist, as if caught in a gigantic spiderweb. The henge was quiet—if one didn't count the single campfire on the stone circle's far side, from where acrid smoke drifted and a lonely drum thumped away like the beating of an unseen heart. The woods hushed and sighed in the rising breeze, which hustled chill tendrils up the sleeves of my fleece jacket.

To my left loomed the solid mass of the horsehead stone. Iris Greene, if that *was* her name, never specified where she wanted to meet me. Since the rock had featured in the counter-curse, I reckoned it made for a suitable meeting place with the enemy. Thankfully, whoever guarded that fire was far enough away, otherwise they might have scratched their heads over a woman cowering next to a storm lantern, guarding a suitcase and a gym bag.

But it was getting late, and Greene never showed. When the

bell in the church tower released a series of melancholic bongs into the gathering darkness, announcing midnight, I'd held my breath—for nothing. Now it was already half past the hour, and still the killer hadn't shown her sneering face.

An invisible hand grabbed my heart and twisted. What if Greene didn't come? What if she kept my Petty? Jenna insisted a familiar was always tied to their witchy creator, but would the killer know? If she didn't...no, my mind shied away from the vision.

Instead, I focused on drawing in air on the count of four, releasing my breath on another count of four. The exercise was supposed to calm my mind and push out the clutter. The guru on the yoga program I once watched said something like, "Don't think, just be."

With so much heartache weighing me down, simply being was a tall order.

Breathe in. Breathe out. Everything will be all right. Don't think; listen to the woods and the beating of the drum. Lose yourself in the here and now...

If only Chris were here. He wanted to come, wanted to help, but we'd agreed it was too risky. Spurning Greene's instructions by bringing along my secret weapon was bad enough. Oh blast, how was I supposed to concentrate when—

Crack. A twig broke.

I jumped from my crouch.

Someone chuckled. "I've never been much of a Girl Scout."

Iris Shuttlecock...no Greene, peeled from the shadow cast by a short, squat standing stone. She carried a mid-sized cardboard box, which quivered and rustled.

A painful breath rushed from my throat. Petty was alive.

Willing myself to remain calm, I pointed at the suitcase. "Here's your stuff. Linda says not to worry about the bill."

"Kind of her, for sure. What sort of innkeeper is she, sticking her nose into her guest's belongings? Where's my diary?"

I hooked a thumb over my shoulder. "Gym bag. Behind me.

You'll get it once you've handed over Petty."

Greene—I'd never call her Iris again—barked a laugh. "Guess what? We'll do it the other way around. You're lucky the stupid thing is no good to me, otherwise you wouldn't get her back at all."

So, she knew.

Fresh pain sliced my heart. Since Petty meant nothing to the woman, my familiar was unlikely to survive this encounter. Most likely, Greene had jinxed the box or done something just as rotten.

I needed to distract the woman to keep her talking. "What triggered you?" I forced an even tone into my voice as if I couldn't care less. "I'm referring to the murders. Why now? Why these people?"

Greene placed the box on the ground. It wobbled, and a whooshing noise drifted into my ear. The pepper scent of Petty's anger followed suit.

Wait, I'll get you out of there.

The box stilled.

The killer didn't seem to notice. She rammed her fists into her waistline, giving a good impression of hurt innocence. In the half-light it was impossible to read her face, but the laughter, mocking and sharp, carried clearly enough. "Gloria shouldn't have threatened me. Emma was even more pathetic than you. Bleat, bleat, bleat. Wanna do magic, wanna do magic. If she ever had magic, she lost it by the wayside, together with her youth. Though that's not why I ended her."

"She knew too much."

"Yes."

"What about Monsieur Poussin?"

"It was an accident." Greene fell silent.

Emma Bingham had said the same.

"I arranged to meet Poussin and summoned Loverboy to the beach. The idea was to have Loverboy bop him one and get arrested."

"And the bucket hat?"

Greene tapped her nose. "Insurance, in case someone saw me. It fell out of her pocket when I handed over the box of potpourri and the envelopes. She never noticed."

"Afterward, when you'd already kidnapped the poor woman, you hid the hat in Emma's apartment. Let me guess—to implicate her?"

"Yes, clever, eh? Seems to have worked miracles."

Her words stabbed my chest. Did we have a traitor in our midst? "How would you know?"

"Your village grapevine is quite something, luv. They know more about the investigation than the cops. Do you now want to hear the story or not?"

"Yes," I squeezed out from between clenched teeth.

"Attagirl. I hired the private dick to find dirt on your sweetheart, to get him into trouble and you away from Avebury. But that bastard had the audacity to say the job was pointless. Your lover was a hacker, yes, but he was being paid for it and everything was aboveboard, blah, blah, blah. He, Poussin, had the folder to prove it. Hah. Went to the office afterward, removed a few sheets and, hey presto, one perfect motive."

Her voice had risen in decibels. "Who did that guy think he was? My cause is just, and he was nothing but a stupid, blundering...anyway, I shoved him. He tumbled down the steps and smashed his dumb head."

Some shove. Self-control didn't count among the woman's strengths.

More heavy breathing, until she said, "At first, I was shocked. I was hiding next to the shed, working out what to do, when Loverboy showed up. I'd totally forgotten I summoned him. Then I spotted this old biddy leaning out of a brightly lit window and talking into a phone. Straightaway, I knew all would be well. And I was right. Poussin's death did wonders for my cause. It was a sign, showing me I must persevere." The shreds of whatever morals the woman might once have possessed tinged her comment with the faintest trace of doubt.

I read somewhere that after the first killing, the next murder became a lot easier. While Greene might not have intended to kill Poussin, his death certainly fit her purpose and provided a lethal inspiration that cost Gloria and Emma their lives.

I sneaked a furtive glance at the stones behind Greene's back. Her focus was on me, Petty's box forgotten. Now was the perfect time for my accomplice to act. What were we waiting for? "You keep talking about a cause. I don't get it."

She licked her lips. "For a supposedly clever person, you really are an ignoramus. Everything started back in the seventeenth century when my stupid great-great-etcetera granddaddy took a shine to that Lily bitch. He wasn't even a young man. Should've known better."

Oh, well done, Linda. Bang on target.

The shadows behind Greene shifted, and a pale face blipped in the darkness. An equally pale hand gave me a thumbs-up.

About time. "He stayed to look after his love?"

"Use your brain. What would she need protection from? No, she had a thing going for the numbskull Reds. Low-skylles Reds. Bah. Soft-hearted little fool. He was no better. Dorks, the two of them. They could have left with the other Whites. Instead, they stayed."

The rushing of the woods segued into a storm. Ice, not blood, flowed in my veins, dampening the *whump-whump* of my heart. Numbness spread in my head until a thought hit like a hurricane.

Lily Coldron had been a White. Not a red witch at all, but a White. Someone who could teleport, levitate, break the world from its angles—

What about Peter Greene? Was he my—and Daisy's—ancestor too? Double the genes and double the trouble? What did that make us?

Who was I?

Like dark eyes glowing from inside a cowl, the answer

ghosted into my mind.

I'm a White.

Not a Red, I was a White, the stuff of nightmares. Feared even by Ignatius's monstrous ancestors.

Giddiness rushed to my head. Control. Absolute power. A scary magical superstar, it was all mine to command.

Some star. And you're control-freaking enough as it is.

My inner voice could be a right bastard, but right now, the sniping was welcome. Whatever I might be, I was still myself. And somewhere deep inside I must have known, or why else didn't I keel over with shock?

Greene tilted her head. "You weren't aware of who you are? Geez, you're pathetic."

A fresh wave of anger scalded my veins. I balled my fists and narrowed my eyes, but that made my foe even harder to spot, so I opened them again.

Perfect timing. Tendrils of a grayish-green smoke snaked from Greene's hands, curling and twisting. Oily slick in the light of my lantern, the immaterial creepers reached for me.

Skylles stop her!

A moist, green banging noise rolled through my head. The ground shook. The grass flattened as if afraid of the elements.

Busy fighting for her balance, Greene never noticed my accomplice poking at Petty's box with a fruit picker.

It was my job to keep the woman preoccupied. "Tell me. What made you act now of all times?" When my gaze fell on the horsehead stone, the answer flashed in bright neon letters. "You found out about the magical plaques. To go through the henge and join the other Whites, you'll need them as an open sesame."

Greene, once more steady on her feet, clapped slowly. "Good girl. During your latest stupid sleuthing stunt, you kept referring to Neolithic plaques. Until then, we only suspected you Coldrons had them."

I ground my teeth. Secure in knowing that—apart from Bob Ignatius—no one knew our plaques doubled up as magical keys.

I'd used them as a cover story in my last murder case.

Behind Greene's back, slim hands worked the fruit picker, drawing the box closer to a squat standing stone.

Note to self—Even Whites don't have eyes in the backs of their heads.

The fruit picker snagged on something. The box fell over.

My heart missed a beat, and I spurted, "Clever of you not to use magic for your murders."

Greene snorted. "Yeah, breaking the first directive can get you killed."

That made her sound a lot more in control than she could have been. She must have been bleeding emotions, otherwise I would never have sensed them in the potpourri she touched. With one eye on the fruit picker, once more groping at the box, I said, "Yet you made plenty of bloopers."

Greene shot up faster than a Weeping Angel. "I never make mistakes. I've a grand plan."

She sounded so Amy Dunne; I had to swallow a hysterical giggle.

Nudged by the fruit picker, Petty's box inched toward safety.

My antagonist's eyes glittered in the moonlight. "Will you still laugh when I tell you my father killed your parents?"

Her words rammed straight into my diaphragm. White noise hissed in my ears. I'd always believed a minion of Ignatius's drove the car that forced my parents off the road and into a standing stone. It appeared I'd been wrong.

Petty's box disappeared, safe in the hands of my secret weapon.

I should be relieved. I wasn't. "Why?" I whispered, my voice hoarse.

"Heavens, he was frustrated, okay? It was so *unfair*. We Greenes should've been guardians of the keys. Like that, he could also stick it to that bastard, Ignatius. He thought Dad worked for him. Did he ever, hah."

Not so wrong, after all.

The hissing in my ears stopped.

Hello? The woman's bastard father killed my parents, implicating Chris's uncle as the instigator, and she was bellyaching about fairness? It was too much. Deep inside my body, something ancient unfurled its thorns, pounced, and crashed through the barriers of civilization. Carried on the scent of leaf mulch and roses, a blazing ball of fury surged until it erupted from my head in a fierce brightness that blinded even me, its source.

Greene shrieked and scuttled aside.

On a wave of perfume, Daisy stepped up and pressed a wooly stocking into my hands.

Cool it, skylles. Stand by.

As my aura faded, the shadows crept back in. One of them moved as Greene peeked around the horsehead stone. "I said, come alone. Your familiar is toast."

"Do you see her anywhere?" Daisy asked. "We knew you wouldn't play fair. So, we didn't either. Oh, by the way, I'm a Coldron too."

"You're no White," Greene snarled. "Not even a full Red. You're second best."

Daisy laughed. "Wrong. Myrtle and Jenna think I'm a fortifier, a catalyst. They seem to be rare, so you might not be aware." Her fingers closed around mine, sharing a reassuring spurt of warmth, just like Jenna did.

Unlike Greene, I would never walk alone.

It was time to end this charade.

In the past, I'd voiced my intentions only in my mind. But this was different. I sucked in a mouthful of breath—and ended up with rose petals on my tongue, which I spat out in a hurry and then tried once more. "Iris Greene, I curse your skylles. You must never use them to evade justice. Nor will you ever mention anything remotely related to magic and the existence of witches, your own included."

My hands, still holding Greene's conjured Christmas

stocking, tingled with sudden warmth. The next moment, the wooly manifestation of the woman's magic inflated and popped in a shower of sparks.

Take that!

Greene, clawing the air in impotent fury, screeched, "I'll kill you. I swear I'll kill the lot of you. I'll curse you right back. I—"

"Just shut your trap, will you?" Once more, Daisy's fingers found mine. Coolness spread from her touch.

"You'll—"

"Go to sleep," I commanded.

Like a puppet whose strings had been cut, Greene collapsed on the grass.

Daisy hugged me tight. "You did it. You really did it."

I hugged her back, and we waltzed over the grass, accompanied by Petty, showering us in sparks. Out of the corner of my eye, I spotted Chris running towards us. "We did it together."

Awareness punched me in the gut. Next stop, confrontation with Sarah.

—

Despite this being the middle of the night, the police presence at the henge had drawn a handful of spectators. They weren't my biggest problem. I'd known Sarah would flip and flip she did.

I held up my hand. "Hold it right there. I understand my meddling"—I curled my fingers into air quotes—"was illegal, but I achieved what you couldn't."

She swung around and banged a fist on the nearest standing stone. "Try me."

"You have to follow the rules and regs," I said. "You need warrants, and that takes time."

"I warned you not to go vigilante. You snubbed my orders and took the law into your own hands."

I fought the urge to yell right back. "I did nothing of the sort.

Instead, I secured reinforcements and called you in as soon as I could."

"You should never have gone against her, reinforcements or not," she stormed. "Don't you trust me?"

"It's that history thing. Greene's family hates my family for something that happened way back in time. It was personal, which is why I needed to talk to her." I bit my lip. I mustn't continue, couldn't share the death of my parents. It would take a while until I'd wrap my mind around that one.

Sarah, still facing the standing stone, spoke over her shoulder. "Soap opera much? You don't expect me to believe this gibberish."

"It's the truth. Ask her." Easy for me to say this, since my curse would prevent Greene from compromising us. For once, I didn't doubt.

She swung back. "Fat chance. This type of perp clams up and lets the lawyer take over. She's no nutcase. She's... I don't know what the world is coming to."

"Makes two of us. Will my statement be enough?"

"It will help, but a good defense attorney will blow holes in the whole construct. We need something tying the woman to the crimes—those in the UK, I mean. Poussin was most likely an accident." She was referring to Mrs. Bingham's note.

"I hope you find your missing link."

I knew she would, for I'd discovered stray tufts of Mrs. Bingham's hair in Greene's pockets. More importantly, Mrs. Mornings had scratched the hand of her killer.

DNA seldom lied.

Killing people the good old-fashioned way to avoid a lethal magical backlash had backlashed in its own way. Between the evidence, Emma's posthumous accusation, and my observations—admittedly slightly doctored—Greene would stew in the slammer a lot longer than Chris.

"Kind of you." Sarah stepped up and the angry glint in her eyes showed despite the sketchy light. "You confronting our suspect is one thing. It's crazy, it's illegal, but I might've looked the other way. You're an intelligent person with morals, and you care about things like justice. But you're hiding something from me. For one thing, how did you catch her in the first place?"

"A great team. Plus hunches. I'm good at them. You said it yourself."

"I'll believe that when pigs fly. Are you MI5? A spy catcher? A spy, even?"

That was beyond idiotic. "Totally not. I'm just who I am."

Sarah's face stilled. "And that's exactly my problem. I don't know who you are anymore. I let you in on the investigation to help your man. You, however? You're being as sneaky as they come. No idea why. You'll have your reasons, I guess, but unless you're willing to tell me the truth, there's no point in continuing our friendship. Goes both ways, you know? Contact me when you're ready."

She turned and walked away.

When I accepted my skylles, I knew this day would come. That didn't stop me from hurting.

Petty's soothing rose scent and a rustling noise drifted by, followed by the intense fragrance of linen and incense.

Chris.

"Pooh," Daisy said. "That's one tough lady."

"I don't blame her," Chris said, his comment yet another stab to my chest.

"How can I expect her to believe in me when I'm... Did you hear?"

"A White. Jenna told me. It's okay."

"Myr, you totally rock." Daisy would have drawn me to her ample chest, but Chris was faster. His breath was warm on my neck. "Everything'll be okay, you'll see."

I hiccupped a sob into his shoulder. "Define okay. This is so crass. It's all my fault. I got you into danger with Greene. She went after you only because of me. What if—"

"Shhh." He put a finger on my lips. "Don't go there. Apropos going. Still fancy a trip to Carnac to detox?"

I pushed away from him and stared. "After all that happened?"

"I doubt it will happen again. Don't forget, since I took the plane, my car is still there." He grinned.

Over his shoulder, I could see Mrs. Greene being read her rights and guided to the back seat of a Panda car, parked under a lantern.

A split-second before she bowed into the car, our gazes connected.

Her eyes flashed. No one but me noticed, but then the message had been meant for me alone.

This wasn't over.

Acknowledgment

Books never are created in a vacuum. Many people and inspirations play into the mix, and so do other authors through their works. Thanks to my parents, who read stories to me and, once I could read them myself, gave me books, I honed my inner Netflix at an early age. A big thank you to both of you!! Another big thank you is owed to my husband Keith, who patiently beta-reads and provides feedback on my crazy creations all the time. I love you, not only for that.

More thanks go to my writing friends from Wattpad as well as The Scribblers's Society and the Writers on Life groups, here especially Marianne, Lisa, Elizabeth, and Vera. Spread across the globe as we are, we still share the same spirit. Long may you write!

And finally I'd like to thank Susie Brooks, Chief Editor of Literary Wanderlust, for all her advice, support – and great artistical talent. You truly are special!

About the Author

Lina Hansen Lina is the author of the Magical Misfits and the Da Vinci series of paranormal cozy mysteries. She has also written a Romantic Comedy, "Spirits of Gascony."

In a previous life, Lina was a freelance travel journalist, teacher, bellydancer, postal clerk and science communication specialist. Not finding enough of what she wanted to read, she set out to write the stories she loves— cozy and romantic mysteries with a dollop of humour and a magical twist. After living and working in the UK, Lina, her husband, and their feline companion now share a home in the foothills of Castle Frankenstein, where she's still writing.

Lina's novels won several awards – two Watty Awards, a First Prize in the 2020 CIBA Awards, a silver book award from Literary Titan, and a five-star recommendation from Indies Today among them.

More information can be found in Lina's blog:
www.linahansenauthor.com
or on Bluesky @linahansenauthor@bsky.social